Introduction to
GEOLOGICAL STRUCTURES

Introduction to GEOLOGICAL STRUCTURES

Amer Ismail
Giuliano Camilletti

Kruger Brentt
Publishers
2025

Kruger Brentt Publishers UK. LTD.
Company Number 9728962

Regd. Office: 68 St Margarets Road, Edgware, Middlesex HA8 9UU

© 2025 AUTHORS

ISBN: 978-1-78715-365-3

For information on all our publications visit our website at http://krugerbrentt.com/

PREFACE

"Introduction to Geological Structures" embarks on a fundamental exploration of the structural features and processes that shape the Earth's crust, offering readers a comprehensive understanding of the principles, methods, and applications of structural geology. This book serves as an essential resource for students, researchers, educators, and professionals seeking to unravel the mysteries of Earth's geological structures and their significance in geoscience and engineering.

Geological structures provide valuable insights into the deformational history, tectonic evolution, and mechanical behavior of Earth's crust, playing a crucial role in the exploration and exploitation of natural resources, the assessment of geohazards, and the understanding of Earth's dynamic processes. From folds and faults to joints and fractures, the study of geological structures encompasses a wide range of features and phenomena that reveal the complex interplay of forces acting within the Earth's lithosphere.

The primary objective of this book is to provide readers with a comprehensive introduction to geological structures, covering key topics such as rock deformation, stress and strain analysis, structural mapping, fault and fracture analysis, and the interpretation of geological maps and cross-sections. Each chapter offers insights, theoretical frameworks, practical techniques, and real-world examples to help readers develop a deep understanding of structural geology concepts and methods.

Contributions from leading experts in the field explore various aspects of geological structures, including the classification and nomenclature of structural features, the mechanics of rock deformation, the role of structural controls in ore deposits and hydrocarbon reservoirs, and the application of structural geology in geotechnical engineering, environmental geology, and natural resource exploration.

Special emphasis is placed on the interdisciplinary nature of structural geology, with discussions on its integration with other branches of geoscience such as sedimentology, stratigraphy, geomorphology, and geophysics. By fostering a holistic understanding of geological structures and their implications, readers can gain valuable insights into Earth's dynamic processes and the ways in which they shape our planet's surface and subsurface.

We extend our sincere gratitude to the contributors whose expertise and dedication have enriched this book. It is our hope that "Introduction to Geological Structures" will serve as a valuable resource and inspiration for students, researchers, and practitioners in the fields of geology, earth sciences, and related disciplines, empowering them to unravel the mysteries of Earth's geological history and contribute to the advancement of knowledge in the field of structural geology.

Amer Ismail

Giuliano Camilletti

CONTENTS

1 Chapter | INTRODUCTION TO GEOLOGICAL STRUCTURES

STRUCTURAL GEOLOGY: AN OVERVIEW

Should the Earth's crust be entirely uniform and homogeneous, discerning any aspects of its history would pose significant challenges. Fortunately, the Earth's crust comprises various structural formations. Structures represent changes in the characteristics of the Earth's crust.

Those variations may be:

- ⊙ Spatial variations: the rocks of the Earth's crust vary from place to place, either on the surface or below; or

- ⊙ Directional variations: rocks look different when viewed from different directions.

For example, where one type of rock contacts another, there is a **geological boundary**, a type of structure. Geological boundaries include:

- ⊙ faults
- ⊙ bedding planes
- ⊙ the edges of igneous intrusions (intrusive contacts)
- ⊙ ancient erosion surfaces (unconformities)

You should have heard about all these types of boundary in your introductory courses. All these boundaries tell you something about the geological history of the area where they are found.

Without examining boundaries, one may discern structure inside a rock unit: the characteristics of several rocks fluctuate with orientation due to the alignment of mineral grains; this phenomenon is referred to as the rock's fabric, a distinct sort of structure.

Figure 1. Mapping a geological boundary

Figure 2. Microscopic thin section of a rock with fabric (field of view 5 mm). The small dark minerals with a strong alignment are biotite mica. Larger grey minerals are mostly the aluminum-rich minerals staurolite and garnet. Light minerals are quartz and feldspar.

Such formations provide significant insights into Earth's history and are essential for those pursuing resources like water, petroleum, and minerals. Certain geological structures developed concurrently with the rocks in which they are located. These constitute fundamental structures. Primary structures encompass beds and laminae in sedimentary rocks such as sandstone and shale, as well as lava pillows in extrusive igneous rocks like basalt. This introduction will address significant forms of primary structures, particularly those crucial for understanding Earth's history, which are typically explored in courses focused on the development of distinct rock types. Numerous formations occur long subsequent to the rocks in which they are located. These represent secondary structures. Secondary structures encompass folds, fractures, foliations in metamorphic rocks, and other more characteristics. The majority of secondary structures result from deformation, which involves the relative movement of crustal components. Structural geology primarily focuses on secondary structures, thereby concentrating on the deformation of the Earth. Tectonics is intrinsically linked to structural geology. Tectonics first denoted the mathematical and geometrical characterization of geological structures at relatively small sizes. In the 1960s, it was discovered that extensive movements of the Earth's outer layer (the lithosphere) could be elucidated by straightforward mathematical and geometrical methods, leading to the emergence of plate tectonics. Since that time, the term tectonics has predominantly denoted the examination of extensive lithospheric processes and the resultant structures.

STRUCTURAL ANALYSIS

The Earth's crust contains structures almost everywhere, and the aims of structural geology are to document and understand these structures. In general, work in structural geology is targeted at three different aims, or levels of understanding.

- **Descriptive** or **Geometric** analysis – what are the positions, orientations, sizes and shapes of structures that exist in the Earth's crust at the present day?

- **Kinematic** analysis – what changes in position, orientation, size, and shape occurred between the formation of the rocks and their present-day configuration? Together, these changes are called **deformation**. Changes in size and shape are called **strain**; strain analysis is a special part of kinematic analysis.

- **Dynamic** analysis – what **forces** operated and how much **energy** was required to deform the rocks into their present configuration? Most often in dynamic analysis we are interested in how concentrated the forces were. **Stress**, or force per unit area, is a common measure of force concentration used in dynamic analysis.

It is essential to maintain these three as separate entities.Primarily, ensure that you can articulate the structures before attempting to ascertain the movements involved, and refrain from making assumptions regarding force or stress without first comprehending both the geometry and kinematics of the scenario.This book will primarily concentrate on the descriptive or geometric purpose, serving as a basis for deeper comprehension.Upon comprehensive description of structures, you will be equipped to advance to kinematic and occasionally dynamic conclusions.

SCALE

Structural geologists look at structures at a variety of scales, ranging from features that affect only a few atoms within mineral grains, to structures that cross whole continents. It's convenient to recognize three different scales of observation.

- ⊙ **Microscopic** structures are those that require optical assistance to make them visible.

- ⊙ **Mesoscopic**, or **outcrop-scale** structures are visible in one view at the Earth's surface without optical assistance.

- ⊙ **Macroscopic, or map-scale** structures are too big to see in one view. They must be mapped to make them visible, or imaged from an aircraft or a satellite.

GEOLOGICAL MAPS: GEOMETRY OF ROCK STRUCTURES

A geological map serves as a potent illustration of the geometry of rock formations. Geological maps are produced by visiting outcrops during fieldwork, when they are described and documented on a topographic base map. The outcome is a geological map documenting the observed rock kinds and features. In several regions, there will be intervals between the visible outcrops, where the bedrock is concealed by soil, vegetation, or various forms of overburden. Creating a geological map requires interpretation to complete the regions between the outcrops. In many instances, a comprehension of geological processes is necessary to formulate an interpretation. Figure 3a presents an outcrop map, while Figure 3b illustrates a geological map created with limited comprehension of geological processes. While it addresses the observations in a rudimentary manner, its accuracy is questionable. Figure 3c represents a more plausible view, informed by our understanding of geological processes. It is important to acknowledge that this second version yields certain kinematic interpretations. It may be inferred that parallel units B, C, and D likely represent sedimentary layers, while unit A may be a newer intrusion due to its cross-cutting relationship with them.

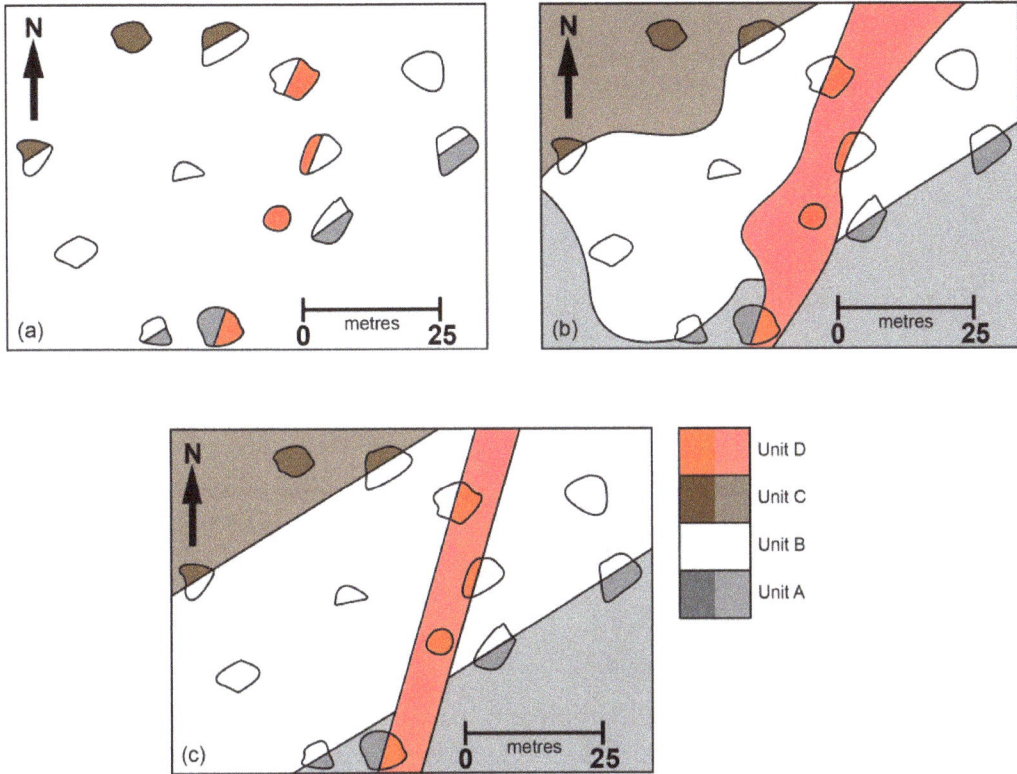

Figure 3. (a) An outcrop map with (b) an unlikely interpretation and (c) a more likely interpretation, producing a reasonable geological map.

2
Chapter

INTRODUCTION TO ORIENTATION OF STRUCTURES

LINES AND PLANES

Linear and planar features in geology

Almost all work on geologic structures is concerned in one way or another with **lines** and **planes**.

The following are examples of linear features that one might observe in rocks, together with some kinematic deductions from them:

- ⊙ **glacial striae** (which reveal the direction of ice movement);
- ⊙ the **fabric** or **lineation** produced by alignment of amphiboles seen in metamorphic rocks (which reveal the direction of stretching acquired during deformation);
- ⊙ and the **alignment** of **elongate clasts** or **fossil shells** in sedimentary rocks (which reveals current direction).

Examples of planar features include:

- ⊙ tabular igneous intrusive bodies such as **dykes** and **sills**;
- ⊙ **bedding planes** in sedimentary rocks;
- ⊙ the **fabric** or **foliation** produced by alignment of sheet silicate minerals such as mica in metamorphic rocks, which reveals the direction of flattening during deformation;
- ⊙ **joints** and **faults** produced by the failure of rocks in response to stress (and which therefore reveal the orientation of stress at some time in the past).

Observe that while numerous preceding descriptive observations yield kinematic inferences, alone the last observation permits dynamic conclusions.

BEARINGS

To characterize nearly any structure, it is essential to address its orientation (sometimes referred to as its attitude): Does it traverse north-south, east-west, or in an intermediate direction? A direction with relation to north is referred to as a bearing. In the majority of geological studies, bearings are designated as azimuths. An azimuth is a bearing measured in a clockwise direction from true north.

An azimuth of 000° represents north, 087° is just a shade north of east, 225° represents southwest, and 315° represents NW.

Figure 1: Compass used to measure an azimuth – in this case the strike of a bedding plane.

Notice that it is best to use a three digit number for azimuths. This helps to avoid confusion with inclinations (below). The degree symbol is often omitted when recording large numbers of azimuths.

Confusingly, there are other methods of specifying an azimuth. In the United States, bearings are often specified using **quadrants**.

In the quadrants method of measuring bearings, angles are measured starting at either due north or due south (whichever is closest), and measured by counting degrees toward the east or west.

Here are the four azimuths above, converted to the quadrants representation:

000°	N00E
087°	N87E
225°	S45W
315°	N45W

Because it is more confusing, especially when doing calculations, we will not use the quadrants method much in this manual. However, you need to be prepared to understand measurements recorded as quadrants, especially when reading books and geologic reports published in the U.S.

Azimuths are typically measured with a **compass**, which uses the Earth's magnetic field as a reference direction. In most parts of the Earth, the magnetic field is not aligned exactly north-south.

The **magnetic declination** is the azimuth of the Earth's magnetic field.

Magnetic declination varies from place to place and varies slowly over time. Currently (2020) the declination in Edmonton is about 014°.

Most geological compasses have a mechanism for compensating for declination. Of course, the compass must be adjusted for the particular area in which you are working.

INCLINATIONS

Another type of measurement is often used in structural geology:

An **inclination** *is an angle of slope measured downward relative to horizontal.*

Figure 2: Compass-clinometer used to measure an inclination – in this case the dip of a bedding plane.

A horizontal line has an inclination of 00°, and a vertical one is inclined 90°. Always use two digits for inclination, to distinguish inclinations from azimuths (three digits).

Inclinations are measured using a device called a **clinometer** or **inclinometer**. Geological compasses typically have a built-in clinometer, so one instrument can be used for measuring both types of angle. However, you must hold the compass differently in each case:

To measure an **azimuth** *precisely, using the Earth's* **magnetic field***, you must hold the compass* horizontal;

To measure an **inclination***, you are using the Earth's* **gravity field***, and the compass must be held in a* vertical plane.

ORIENTATION OF A LINE

To specify the orientation of a line requires two measurements, called **plunge** and **trend**:

The **plunge** *of a line is its inclination, measured downward relative to horizontal;*

The **trend** *of a line is its azimuth, measured in the direction of plunge.*

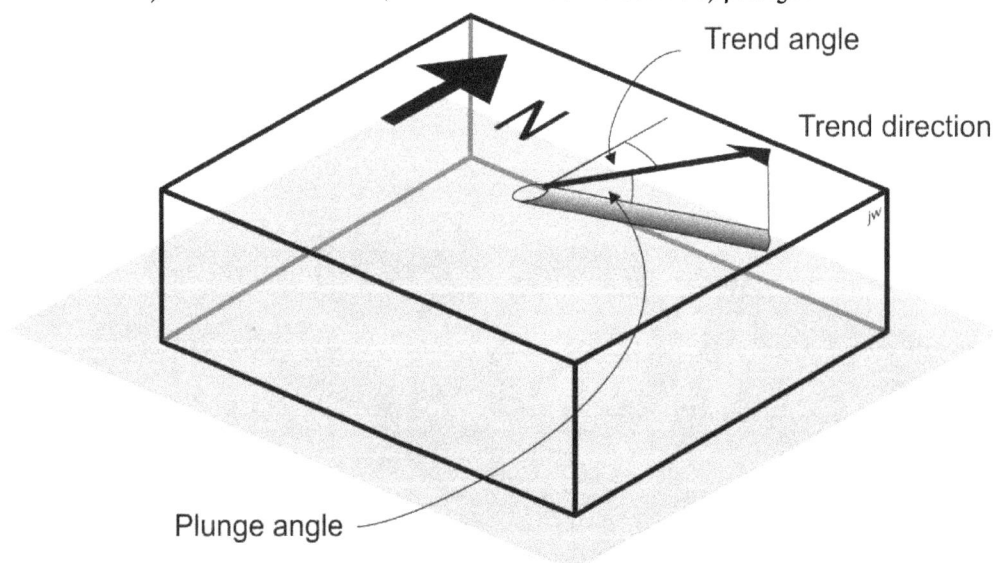

Figure 3: Trend and plunge of a linear geological feature.

So, a line with plunge 07 and trend 007 slopes downward very gently in a direction just east of north. 227-87 specifies a line that plunges very steeply towards the SW.

There are several different conventions for writing plunge and trend measurements: some geologists write the plunge first and some write it second. The best way to keep things clear is to always use three digits for the trend and two for the plunge.

In addition, it's sometimes helpful to specify the compass direction, just as a check, e.g.

025-37 NE

ORIENTATION OF A PLANE

To specify the orientation of a plane, we also need two measurements, an azimuth and an inclination. The **dip** of a plane is its inclination. It's important when measuring dip to measure the steepest possible slope in the plane. If you are in doubt, imagine water running down the surface; it will take the steepest path, in the direction of dip.

The dip *of a plane is the inclination of the steepest line in the plane.*

The azimuth of a plane is a bit more complicated. There are several different directions that we might measure. If we measure the direction in which the plane slopes downhill, then we are measuring **dip direction.**

The **dip direction** *of the plane is the azimuth of the steepest line in the plane.*

However, dip direction is not easy to measure accurately with many compasses, because the slope of the plane varies rather gradually on either side of the dip direction. For this reason, many geologists prefer to measure the strike, which refers to the direction of a horizontal line drawn on the surface.

The **strike** *of a plane is the azimuth of a horizontal line that lies in the plane.*

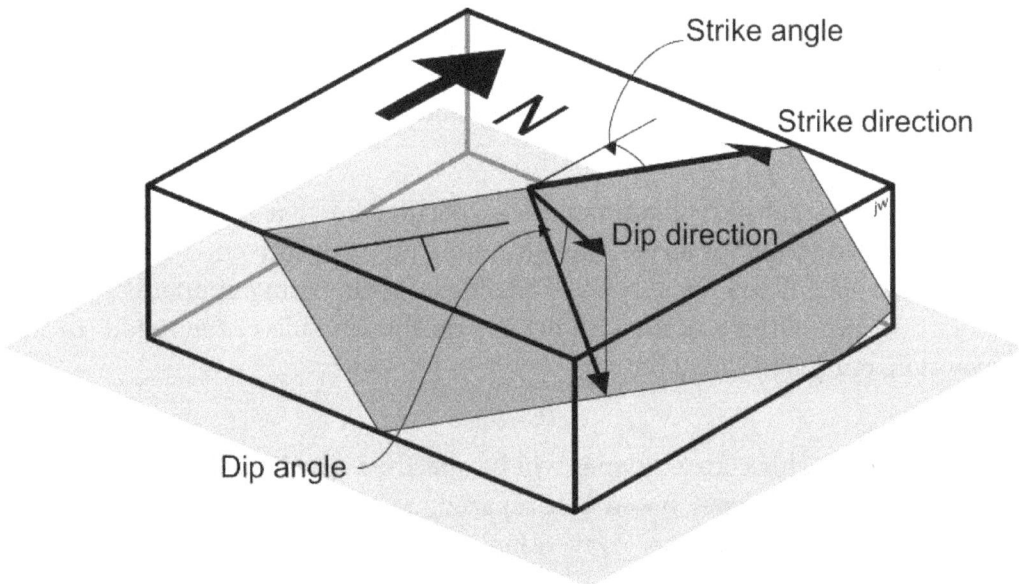

Figure 4. Strike, dip, and dip-direction of a plane.

There are two directions in which we could measure the strike, 180° apart! The dip direction is clockwise from one, and counterclockwise from the other. In most Canadian geological field work, the **right-hand rule** ('**RHR**') is used to avoid this ambiguity.

Right-hand rule: *When you are facing in the strike direction, the plane dips downward to your right.*

An equivalent statement is that strike is always 90° counter-clockwise from the dip direction.

It's a good idea to add a rough compass direction to the dip measurement, just as a check that right-hand rule measurement has been done correctly. For example:

345/45 NE

specifies a plane that dips at 45° with strike roughly NNW. The dip direction is clockwise from the strike, so the dip direction is ENE – but 'NE' indicates that we have the direction right.

Other conventions for defining the orientation of a plane

Unfortunately, there are several other conventions to resolve strike ambiguity.

Some geologists prefer to record whichever strike direction is less than 180, and use letters (e.g. 'NE') to resolve the ambiguity. In this convention ('strike, dip, alphabetic dip direction') the above measurement would be written:

165/45 NE

Other geologists prefer to record dip direction and dip. In the 'dip-direction, dip' (DDD) convention, the above measurement would be written:

075,45

In the UK the strike has sometimes been specified so that the dip direction is counterclockwise from the strike, though confusingly this convention is also called 'right-hand rule'. If you want to know the logic for this convention, ask a British geologist! (It has nothing to do with driving on the left side of the road.) In this convention, our plane would be:

165/45

In most work for this course, planes will be specified using the (Canadian) right-hand rule. However, you should be prepared, as geologists, to work with data collected using any of the other conventions.

RELATIONSHIP OF LINES TO PLANES

It is frequently feasible to measure many linear and planar features at a single outcrop. Occasionally, unique relationships exist among these structures. The subsequent sections delineate several of these relationships.

Intersecting planes

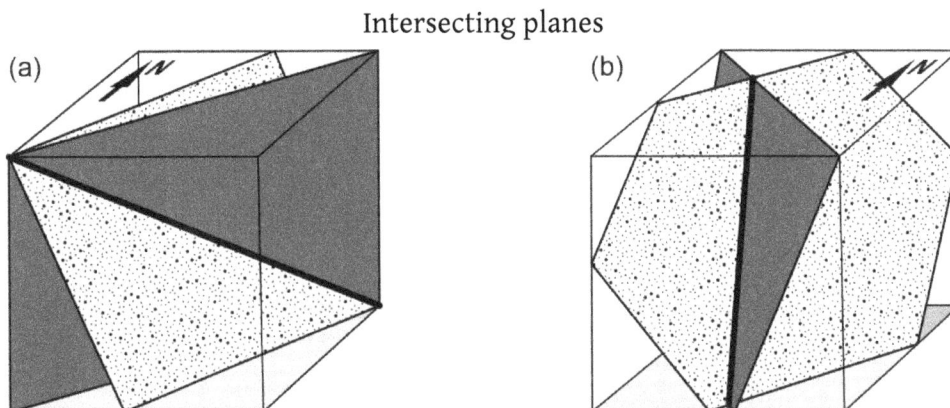

Figure 5: Two examples of intersecting geologic planes. In (a) a dipping plane (stippled) intersects with a vertical plane (shaded) to produce a plunging line of intersection. In (b) neither plane is vertical.

If two planar structures possess distinct orientations, they will collide in three-dimensional space. The convergence of two nonparallel planes forms a line (Fig 5).The orientation of the intersecting line is solely determined by the orientations of the two planes.Altering the position of one or both planes while maintaining their orientation will modify the location of the line of intersection, albeit its orientation will remain unchanged.This guidebook presents numerous scenarios involving the intersection of planes. The subsequent points are of great significance:

- ⊙ The intersection of a geological surface with the topographic surface (the ground) is called the **surface trace** or **outcrop trace** (or just **trace**) of that surface. Geological maps are typically divided into areas of different colours (for different rock units) that are bounded by lines; these lines on the map are the traces of the geological surfaces that separate the units.

- ⊙ The intersection of a fault plane with a planar rock unit that the fault displaces produces a line called the **fault cut-off or cutoff**.

- ⊙ The two sides, or limbs, of a fold may intersect on a line called the fold **hinge**.

- ⊙ The truncation, at an unconformity, of an older planar rock unit or surface by a younger one with a different orientation in space produces a line which may be called the **subcrop**, or **subcrop limit**.

LINE THAT LIES IN A PLANE

On any plane, an endless number of lines can be drawn that are parallel to or reside within the plane. Examples include contemporary lineations situated on bedding planes and striations present on the fault planes itself. The alignment of a line within a plane can be defined by rake or pitch. In contrast to an azimuth (measured from north in a horizontal plane) or an inclination (measured from horizontal in a vertical plane), a rake is measured from horizontal in an inclined plane, as seen in Fig. 6. Similar to strike, there are multiple conventions for delineating rake. We recommend measuring the rake of a line from the 'right-hand rule' strike direction, clockwise when looking down on the surface, as an angle between 000° and 180°.

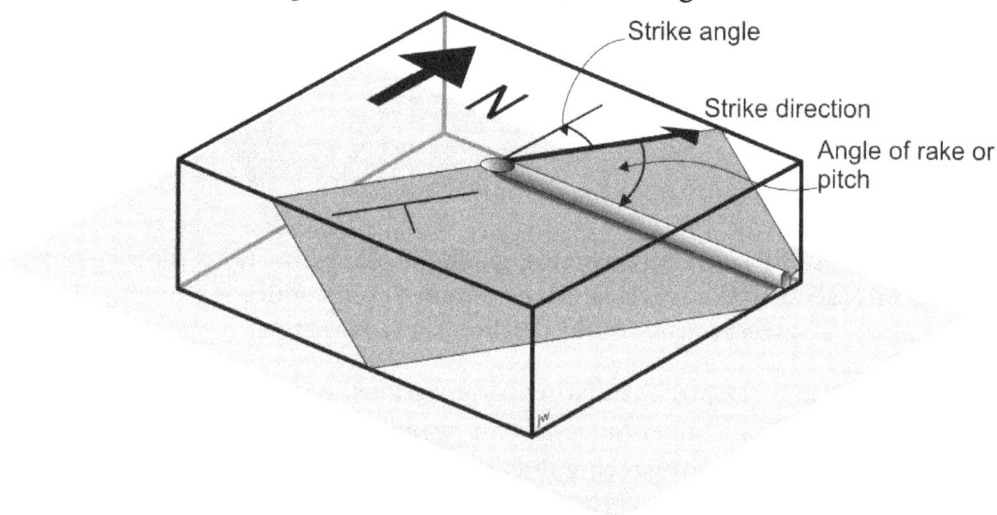

Figure 6: Rake of a linear feature in a planar feature.

For example, a geologist may record a fault surface like this:

Fault plane 075/78 SE; Slickenlines rake 108°

On a vertical plane the rake of a line is the same as its plunge. On all other planes, rake ≥ plunge.

Remember: it only makes sense to measure a rake when a line lies in a plane.

Pole to a plane

There's also an infinite number of other lines are *not* parallel to any given plane (they may **pierce** the plane). One special line is perpendicular to any given plane: it's sometimes called the **pole** to the plane. We will meet poles to planes in a later section of the course.

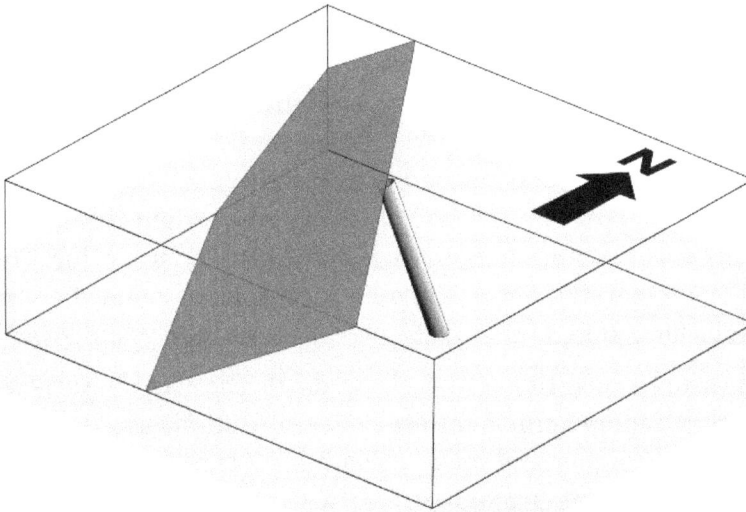

Figure 7: Orientation of a plane and its pole.

CONTOURS

What are contours?

Contours are curving lines on a map that are widely used in the Earth sciences to show the variation of some quantity over the Earth's surface. You are probably most familiar with **topographic contours** that show the shape of the land surface. However, Earth science uses many other types of contours such as:

⊙ **Magnetic contours**: Variations in the strength of the Earth's magnetic field;

⊙ **Isobars**: Variations in air pressure;

⊙ **Isopachs**: Variations in the thickness of a stratigraphic unit;

⊙ **Structure contours**: Variations in the elevation above sea-level or depth below sea-level of a geological surface.

In each of these cases a numerical quantity, such as the elevation of a surface, varies from place to place, and the contour lines illustrate that spatial variation.

A **contour** *is a curving line on a map that separates higher values of some quantity from lower values.*

A contour can also be thought of as a line connecting points at which the measured quantity has constant value. Each contour line is labelled with this constant value; a map covered with contour lines is a useful expression of the spatial variation of the measured quantity.

(Note: This property is sometimes used as a *definition* of a contour. For example, a topographic contour is sometimes defined '*as a line joining points of equal elevation*'. Although this is a satisfactory definition, it is harder to apply in practice, for two reasons. First, when the data are sparse, for example when working with drilled wells, it may be difficult to find any points of exactly equal elevation; locating such points requires **interpolation**. Second, it is very easy, when **threading** contours, to end up with "lower" points on both sides of the same contour line. This is always wrong! So, it is imperative when drawing a contour to remember that it has a 'high' side and a 'low' side, so that it always separates higher and lower values.)

Often, the measured quantity is the elevation of the Earth's surface, above or below sea level. A **topographic contour** can be considered as a line on the ground separating points of higher and lower elevation. It can also be thought of as the line of intersection of the ground surface with a horizontal plane. Below sea level, contours showing the elevation of the sea floor are known as **bathymetric** contours.

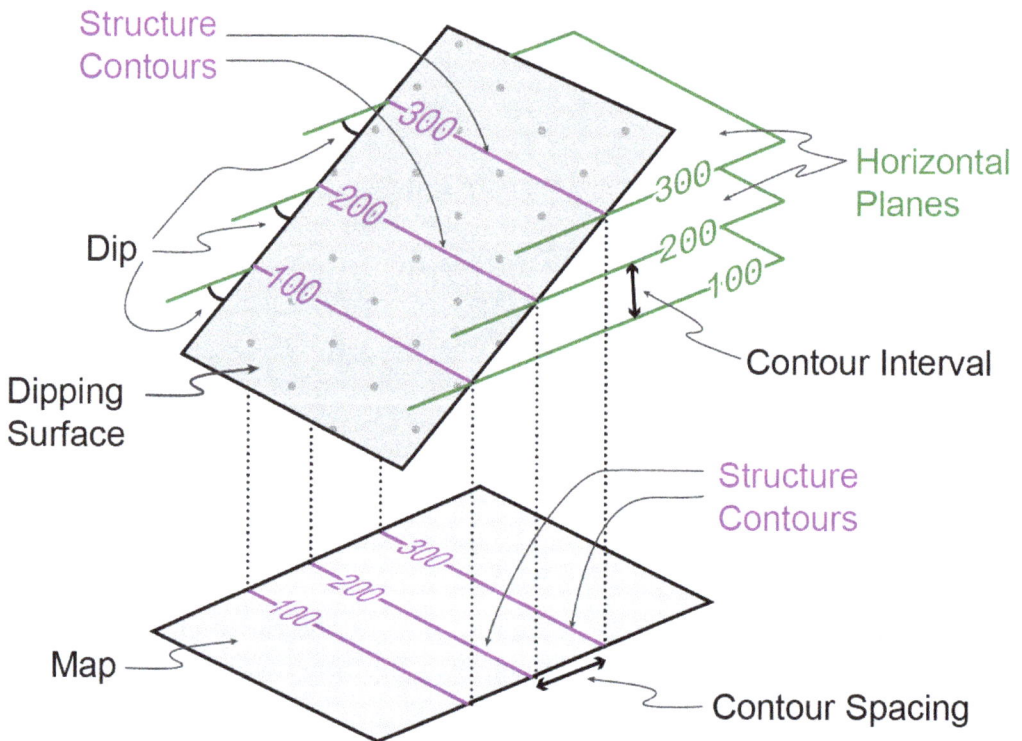

Figure 8. Relationship between dip and contour spacing.

On most topographic maps, topographic contours are separated by a constant interval: for example, contours on a map might be drawn at 310, 320, 330, 340 m etc. The spacing of the contours is called the **contour interval**. In this example the contour interval is 10 m.

A **structure contour** (Fig. 8) is a contour line on a geologic surface, such as the top or bottom of a rock formation, a fault, or an unconformity. Typically, structure contours are drawn on surfaces that are buried underground. However, sometimes it's possible to guess where a geologic surface was before it was eroded away; structure contours are then drawn for this imaginary surface above ground! Just like a topographic contour, a structure contour is the line of intersection of the contoured surface with a horizontal plane.

STRIKE, DIP, AND CONTOURS

Because structure contours are by definition lines of constant elevation, they are parallel to the strike of the geologic surface. They are sometimes called strike lines. So, given a pattern of structure contours it's possible to determine the strike of the surface at any point.

The dip of the surface controls how far apart the contours are. Where a surface dips steeply, the contours are close together; where the surface is near-horizontal the contours are far apart. The horizontal spacing of contours, recorded on the map is called the **contour spacing**. There is a simple relationship between the dip δ of a surface and the spacing of its contours.

$$\tan (\delta) = \text{contour interval} / \text{contour spacing}$$

If a surface is **planar** (i.e. the strike and dip are constant) then the contours will be *parallel, equally spaced, straight lines.* Thus you can readily determine the orientation of a surface from the azimuth and spacing of its structure contours.

Contours and outcrop traces

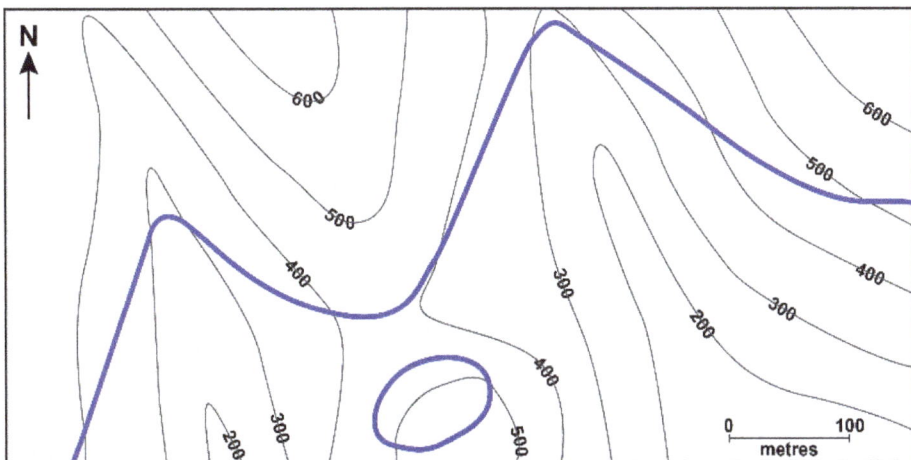

Figure 9. Map showing topographic contours and the outcrop trace of a single geological surface.

On a geological map, a geological surface, such as the boundary between two map units, is represented as a line known as the outcrop trace or topographic trace of that boundary (Fig. 9).The outcrop traces of geological units are generally intricate, curvilinear formations, influenced by both the dip of the geological surface and the complex morphology of the topographic surface.Consequently, in regions with topographic relief, it is frequently feasible to utilize the outcrop trace of a boundary between two rock units to deduce the strike and dip of the units. The exact orientation of a surface can be ascertained from its outcrop trace, as its position and elevation are identifiable at each intersection with a topographic contour line.The junction points can be utilized for delineating structural outlines (Fig. 10).Consequently, the 400 m structural contour is delineated by linking all locations where the outcrop trace intersects the 400 m topographic contour.After several structure contours have been delineated, the surface orientation can be ascertained from the spacing and alignment of the contours.

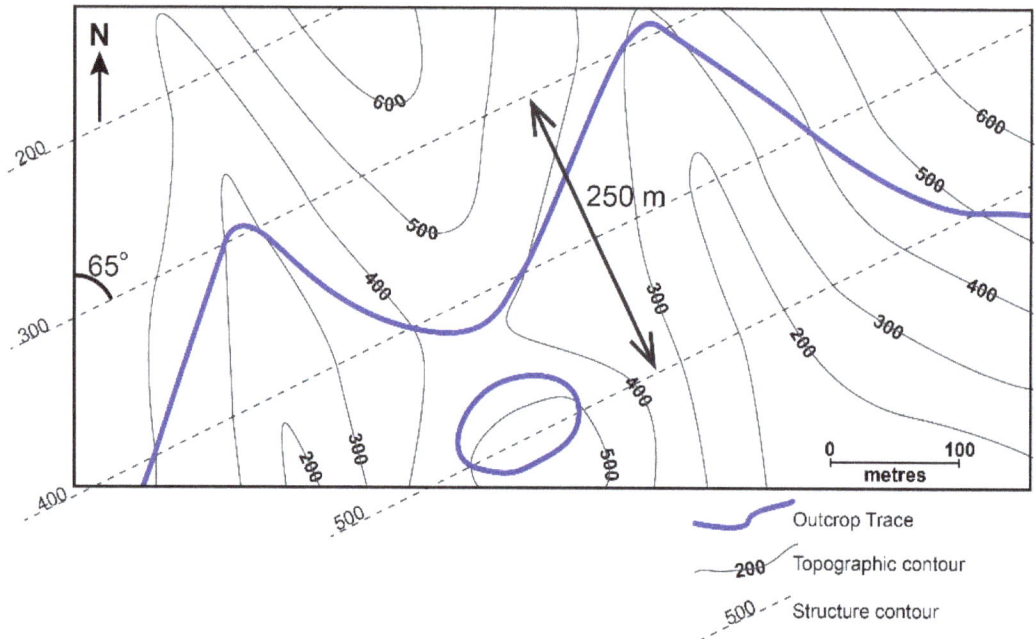

Figure 10. Structure contour construction on the map in Fig. 9. The strike and dip of the surface can be determined from the contour orientation and spacing. In this case, the structure contours are oriented 65° from north, but the numbers on the contours tell us that the surface gets lower towards the NW, so the RHR strike is: 65° + 180° = 245°. The structure contours are 125 m apart and the contour interval is 100 m. Dip = arctan(100/125) = 39°. Therefore the RHR orientation of the surface is: 245/39 NW

Conversely, if the structural contours of a geological surface are established, its trace can be ascertained by linking sites at which the geological and topographic surfaces share identical elevations; that is, the trace joins places where the structural and topographic contours intersect at the same elevation.When the elevation of a structural contour exceeds the topographic elevation, it indicates that the geological surface is "above ground" and has consequently been eroded at that site. Conversely, when the height of a structural contour is lower than the topographic elevation, it indicates that the geological surface lies beneath the earth, or in the subsurface, and can be accessed through excavation or drilling.The outcrop trace of a geological surface can be regarded as a line demarcating an area where the surface exists subterraneously from another area where the surface has been eroded above the current ground level.

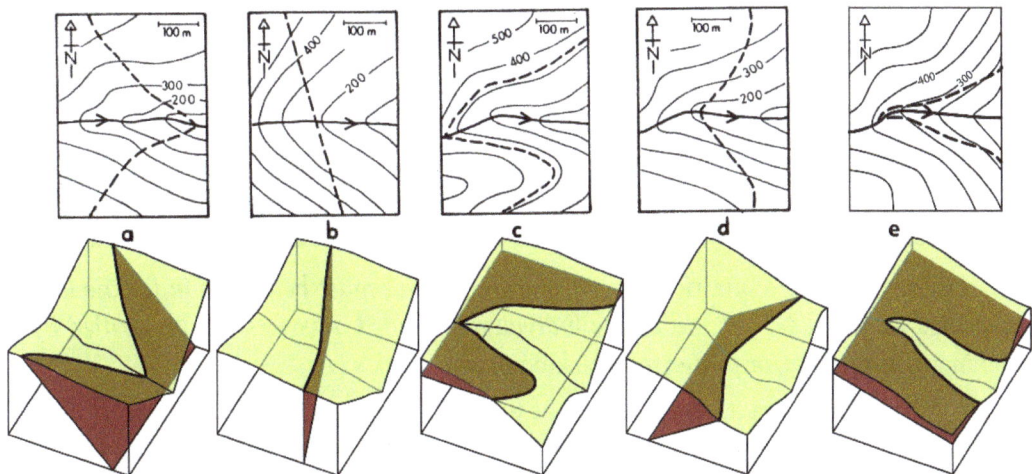

Figure 11. Sketch maps and block diagrams showing the outcrop traces (dashed lines) of geological surfaces of different orientation: (a) Dip to the east; (b) Vertical; (c) Horizontal; (d) Dip to the west; (e) Dip to the east but less steep than valley.

⊙ There are some general considerations when constructing geologic traces (Fig. 11).

⊙ The outcrop trace of a horizontal geological surface is parallel to the topographic contours.

⊙ The outcrop trace of a vertical geological surface is a straight line parallel to the strike; it ignores topographic contours.

⊙ The outcrop traces of dipping surfaces show V-shapes as they cross valleys and ridges; these regions are particularly useful in determining strike and dip.

⊙ In general, the V-shape formed as a trace crosses a river valley points in the direction of dip. (This is known as the "rule of vees".) The only exception occurs when the dip is in the same direction as the slope of the valley, but gentler than the gradient of the river; then the V-shapes point up-dip.

- ◉ For planar surfaces with shallow dip (gentler than the typical hill slopes of topography in the region) the outcrop trace will generally follow topographic contours quite closely, crossing them at widely spaced intervals.

- ◉ In such regions, the relative position of a top or bottom contact of a unit can be inferred from the local topography. For example, if the position of the bottom trace of a unit is known then the top of the unit must be exposed at a higher elevation.

A geologic trace should never cross a topographic contour except where the identical structural and topographic contours intersect.

LAB 1. ORIENTATION OF LINES AND PLANES

Address the questions in any sequence to mitigate traffic congestion surrounding the rock samples. Rock samples may be inaccessible outside laboratory hours.

An asterisk indicates a question for self-marking. These must be passed in for the lab to be verified as complete, but will not be individually graded. Answers will be verified by a teaching assistant and/or posted for checking in next week's lab.

1. * To make sure you are conversant with both the quadrant convention (widely used in the USA) and the azimuth convention (used in Canada and most of the rest of the world) for recording bearings, translate the azimuth convention into the quadrant convention, and vice versa, for the following bearings.

    a) N12E

    b) 298

    c) N62W

    d) S55W

2. * Rock samples containing planar structures are set up in the laboratory. **(a)** Using a compass-clinometer, measure the strike of a planar structure. To do this, hold the compass in a horizontal plane so that the needle swings freely in the Earth's magnetic field. Then place the compass so that its horizontal edge is against the surface, keeping the compass level. Note the reading of the compass needle (your instructors will show you how to read the particular model of compass). **(b)** Now measure the dip. To do this, turn the compass so that it is in a vertical plane (the pendulum or spirit bubble – depending on the type of compass – should swing freely in the Earth's gravity field). Place the compass so that its edge is in contact with the surface along the steepest slope. Read the dip (your instructors will show you how to read the number for the particular model of compass). **(c)** Record the strike (right-hand-rule) and the

dip. **(d)** Also record the dip direction (N, S, E, W) as a check. **(e)** Repeat for the other samples as directed. (Note that the answers you get will probably not be true orientations because the Earth's magnetic field will be distorted by metal objects in the building: in other words, the declination of the Earth's magnetic field is highly variable indoors.)

* When you are done, have a teaching assistant check and initial your answers.

3. * Translate the following orientation measurements from the dip-direction and dip (e.g., 060°, 45°) convention into the North American right hand rule convention, adding an alphabetic dip direction as a check (e.g. 330°/45°NE).

   a) 177°, 13°

   b) 032°, 45°

   c) 287°, 80°

4. * Translate the following orientation measurements from the strike, dip, and alphabetic dip-direction (e.g., 087°/ 21°N) into the North American right hand rule convention.

   a) 087°/21°N

   b) 005°/73°W

   c) 042°/30°SE

5. * Rock samples containing linear structures are set up in the laboratory. **(a)** Using a compass-clinometer, measure the trend of a linear structure. To do this, hold the compass in a horizontal plane so that the needle swings freely in the Earth's magnetic field. Then place the compass so that its horizontal edge is over the plunging line, keeping the compass level. Note the reading of the compass needle (using the same method as before). **(b)** Now measure the plunge. To do this, turn the compass so that it is in a vertical plane (the pendulum or spirit bubble – depending on the type of compass – should swing freely in the Earth's gravity field). Place the compass so that its edge is in contact with the line. Read the plunge (using the same method as for dip). **(c)** Record the plunge and the trend. **(d)** Add the trend direction (N, NE, E...) as a check. **(e)** Repeat for the other sample(s) as directed. (Note that the answers you get will probably not be true orientations because the Earth's magnetic field will be distorted by metal objects in the building: in other words, the declination of the Earth's magnetic field is highly variable indoors.)

   * Have a teaching assistant check and initial your answers.

6. * Topographic and geological surfaces are not always planar. When a surface is curved, the strike and dip vary from place to place. When this happens, **contouring** is a good way to reveal the shape of the surface. Map 1 shows an area of map in which the elevation has been measured at a large number of points. *Place tracing paper over your map.* Using a contour interval of 100 m, thread con-

tours through the measured points. In making your contour map, remember the following points:

- Maps and map scales: You should be able to convert a map scale expressed as a representative fraction (e.g. 1:50,000) to a map scale expressed in metric or imperial length units (2 cm = 1 km) and draw a **scale bar** based on either. You should understand **topographic contours** and should be able to look at a map with topographic contours and identify hills, valleys, and predict which way streams are flowing.

- Your contour map is a hypothesis: it should be the simplest map that is consistent with the data, so your contours should be as smooth as possible; avoid sharp bends and changes in the spacing of contours, unless required by the data;

- Each contour has a high side and a low side. The 200 m contour (for example) separates ground that is above 200 m (on one side) from ground below 200 m (on the other);

- Contours can never branch;

- Rivers flow at the lowest point of a valley and must always flow downhill in the same direction.

  Add arrows to show the direction of flow.

7. Map 2 shows the configuration of the topographic surface and the trace of the **top surface** of a unit of banded iron formation. Determine the orientation of the surface, and shade the area of the map where the iron formation crops out.

8. * Map 3 shows the same area as Map 1. Place the contour map you made in the earlier question over Map 3 to compare the contours. If there are differences, try to evaluate whether the two interpretations are equally valid. Show with dashed lines any places where you think changes to your map are required by the data

9. A thin coal seam was observed at point X on Map 3. Unlike the topographic surface, it is perfectly planar; the orientation is everywhere 010°/14°. Determine the spacing and orientation of the structure contours, and draw this second set of contours on Map 3.

Complete the trace of the coal seam. You may assume that the soil cover is negligible.

3 Chapter

INTRODUCTION TO PRIMARY STRUCTURES

Structural geology primarily concerns the formations that emerged in rocks as a result of deformation caused by tectonic processes.Nevertheless, when detailing buildings, it is typical to encounter formations that emerged concurrently with the rock development.These are referred to as fundamental structures.Understanding these structures is essential for numerous reasons.They can provide insights into the conditions under which the rocks formed and may assist in the investigation of subsequent deformation.In sedimentary rocks, fundamental structures serve as crucial indicators of the formation environment.The examination of sedimentary structures is a primary emphasis in sedimentology courses.This survey will focus on the most significant sedimentary structures, particularly those most beneficial to structural geologists.Occasionally, these formations provide insights about whether the rocks have been inverted since their inception, making it essential for structural geologists, stratigraphers, and paleontologists to identify and comprehend them. In igneous rocks, primary structures are significant as they indicate whether an igneous mass is intrusive or extrusive.Both sedimentary and extrusive igneous rocks are frequently stratified, structured into layers (strata) that were once horizontal.Structural geologists collect data on Earth's history by assessing the orientation and sequence of formation of geological structures. Thick strata are represented on geological maps as discrete entities, referred to as formations, groups, members, etc. Stratigraphy is the study of the organization of strata.

STRATA AND STRATIGRAPHY

Basics

Numerous sedimentary and certain igneous rocks are stratified, consisting of layers (strata) deposited parallel to the Earth's surface (principle of original horizontality), with the oldest layers at the bottom and the youngest at the top (principle of superposition). Stratified units predominate on numerous geological maps.They are typically represented in various colors.

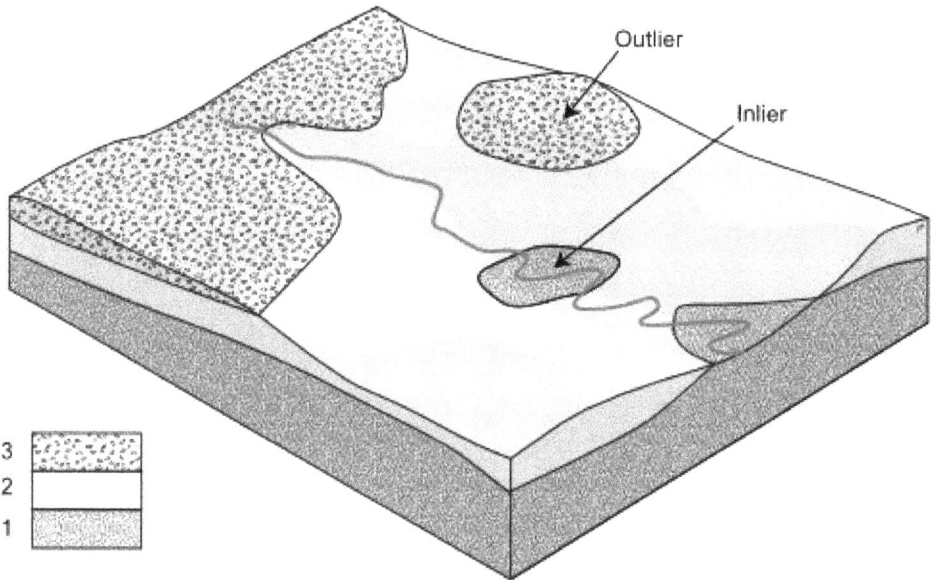

Figure 1. Cross-section and map view showing how the relationship between older and younger units develops for inliers and outliers. Oldest unit is 1, youngest is 3. Inlier of 1 is exposed through overlying unit 2 and thus surrounded by it in map view.

The interplay of topography and geology can occasionally provide intricate map patterns, characterized by areas of one unit encircled by others.An inlier consists of older strata exposed and encircled by younger strata, while an outlier is an exposure or erosional remnant of younger strata entirely enclosed by older strata. Figure 1 illustrates these relationships, depicting an eroded succession of strata; unit 1 is the oldest, while unit 3 is the youngest.

STRATIGRAPHIC UNITS

Formations

The fundamental unit of stratigraphic mapping in layered rocks is the formation. One of the initial tasks in surveying a new region is to delineate formations.

A formation must be:

- Mappable at whatever scale of mapping is commonly practised in a region;
- Defined by characteristics of lithology that allow it to be recognized;
- It must have a type section that exemplifies those characteristics;
- It must be named after a place or geographical feature.

Additional regulations for delineating formations are specified in the North American Stratigraphic Code and the International Stratigraphic Code.

OTHER LITHOSTRATIGRAPHIC UNITS

Formations can be categorized into groups.Subordinate mappable units may occasionally be identified within formations.These are referred to as members. Formations, groups, and members constitute lithostratigraphic units.This indicates that they are determined only by lithological properties; age is excluded from the definition.Calculations of thicknessColumnar representations of measured sections, illustrating the thicknesses of strata, are frequently employed in stratigraphy and sedimentology. When strata are well exposed, one can measure the thickness of each bed using a tape measure. It is frequently essential to measure diagonally across inclined strata; therefore, apparent thicknesses must be adjusted to account for the discrepancy between the measured direction and the direction perpendicular to the strata, known as the pole to bedding. Numerous trigonometric formulas exist to convert true thickness to apparent thickness for various combinations of inclined planes and plunging lines of section.Nonetheless, all represent variations of a singular fundamental formula.True thickness = Apparent thickness × $\cos(\theta)$

where θ is the angle between the line of section and the 'ideal' direction represented by the pole to stratification. This angle can easily be determined from a stereographic projection, which we will cover in the next section.

Unconformities

Unconformities are ancient surfaces of erosion and/or non-deposition that indicate a **gap** or **hiatus** in the stratigraphic record. An unconformity may be represented on a map by a different type of line from that used for other geological contacts; in a cross-section an unconformity is often shown by a wavy or crenulated line.

Subtle unconformities are very important in the analysis of sedimentary successions. A **sequence** is a package of strata bounded *both above and below* by unconformities. It may be that an unconformity is a sequence boundary, but that determination depends on finding another unconformity in the succession, either higher up or lower down stratigraphically.

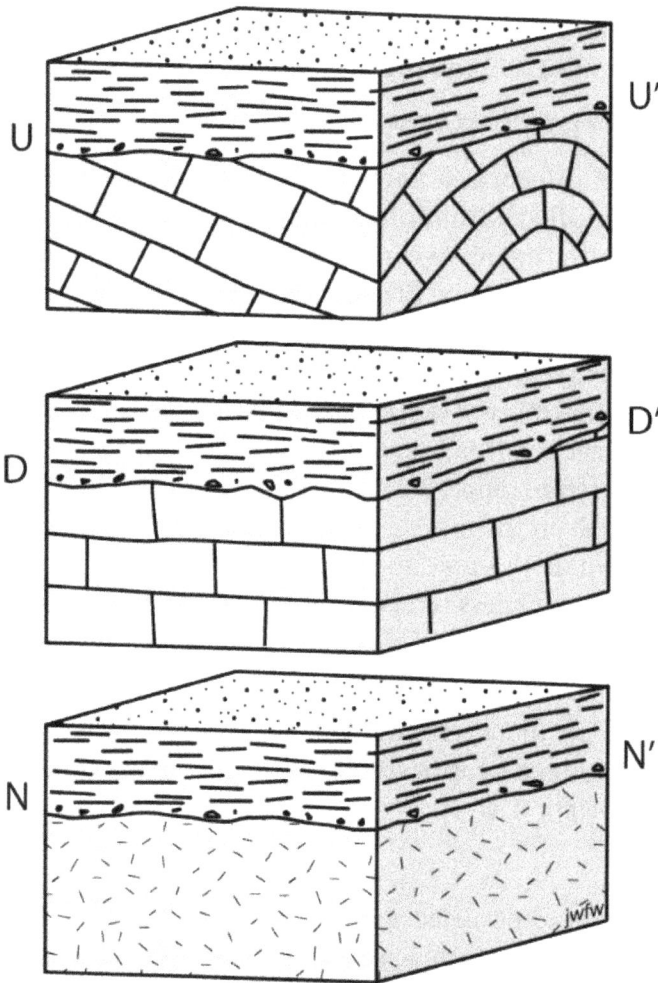

Figure 2. Unconformities. Block diagram of an angular unconformity (UU'), a disconformity (DD'), and a nonconformity (NN').

Angular unconformities

An **angular unconformity** is characterized by an angular **discordance,** a difference of strike or dip or both, between older strata below and younger strata above (Fig. 2a). In the diagram, the younger strata are horizontal. However, subsequent tilting of the entire succession could alter the orientations, but there will still be a discordance between strata above and below the unconformity. Bedding in the younger sequence tends to be parallel to the plane of unconformity, or nearly so.

Structure contours on the younger strata above the unconformity will not have the same orientation and spacing as those on the older strata below; either the strike or the dip, or both, will be different.

Figure 3. Angular unconformity, sandstone and conglomerate (top of cliff) of the Triassic Fundy Group resting on near vertical mudstone and sandstone of the Carboniferous Horton Group, Rainy Cove, Nova Scotia.

Figure 4. Close-up of the same angular unconformity at beach level, Whale Cove, Nova Scotia.

Disconformities

Figure 5. Disconformity. Middle Ordovician limestone of the Table Head Group (darker grey) resting on eroded surface of Lower Ordovician dolostone of the St. George Group (paler grey and brown), Aguathuna Quarry, Port au Port Peninsula, Newfoundland.

A disconformity is a contact between parallel strata with a considerable age difference. What are the methods of recognition?One can identify the erosion surface by examining features such as stream channels, buried soil profiles, and pebbles or conglomerates.If the sole evidence for a gap arises from comprehensive paleontological analysis, the surface is classified as a paraconformity. In both scenarios, the structural outlines for strata above and below the unconformity will be parallel and exhibit identical spacing.

Nonconformities

The term **nonconformity** is used to describe the contact between **younger** sedimentary strata deposited upon an eroded surface of **older** crystalline (igneous or metamorphic) rocks (Fig. 2c) in which distinct layers cannot easily be recognized. (In other words it is impossible to say whether the unconformity is angular or not.)

Figure 6. Nonconformity. Proterozoic Torridon Group resting on Paleoproterozoic gneiss, NW Scotland, UK.

Exercise caution in differentiating a nonconformity from an intrusive encounter. Younger sedimentary rocks were deposited on an earlier incursion at a nonconformity. Sedimentary rocks often contain pebbles of weathered intrusive rock. In an intrusive contact, the intrusion is the younger component. Adjacent older sedimentary rocks may exhibit baking (thermal metamorphism), and the younger intrusion may incorporate fragments (xenoliths) of the older sedimentary rocks. An obtrusive contact, albeit being significantly discordant, does not constitute an unconformity.

Unconformities in geologic maps

Angular unconformities are often identifiable on geological maps. The succession above the unconformity generally exhibits strata that are nearly parallel to the unconformity, whereas the rocks beneah the unconformity are distinctly truncated at the boundary.

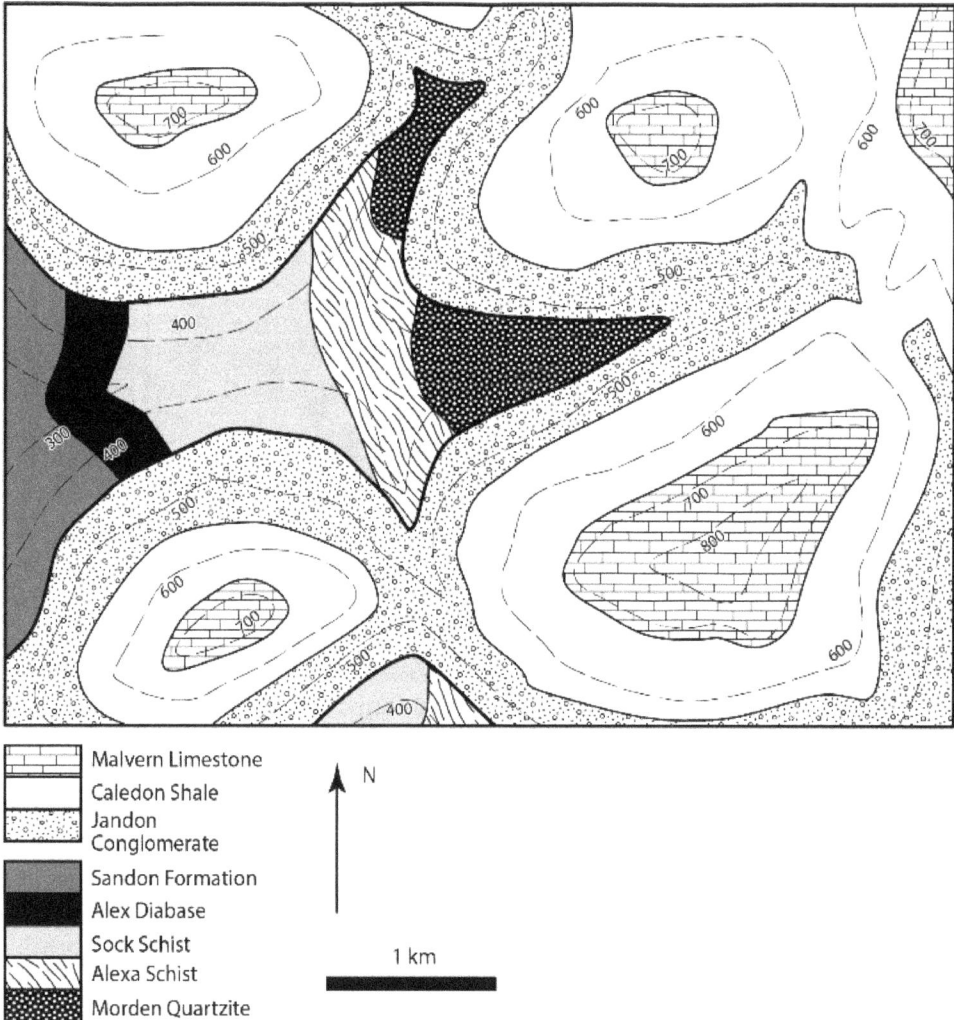

Figure 7. Geological map that includes an unconformity.

Onlap and overstep

In a sedimentary sequence featuring an unconformity, the strata may exhibit either onlap, overstep, or both (Fig. 8).In that section of the succession above an unconformity, younger strata onlap the underlying succession if progressively younger layers spread further geographically over the unconformity surface.The onlap connection typically results from the gradual burial of topography through a process known as transgression, which is defined as the landward migration of the shoreline.Overstep refers to a relationship where strata beneath the unconformity are involved, illustrating how the younger succession is positioned atop various units within the lower succession.In Figure 8, Y and Z onlap the unconformity, whereas Y oversteps from R onto S and T.

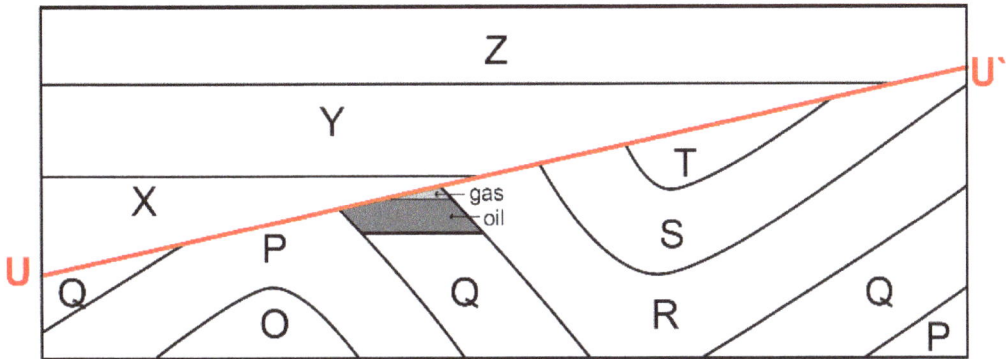

Figure 8. Cross-section of an angular unconformity UU`, showing onlap and overstep. Unit X is impermeable, but unit Q is porous, and contains a petroleum reservoir in a subcrop trap.

Paleogeologic maps and subcrops

Rocks exposed at the current erosion surface are termed outcrops, while those exposed at historical erosion surfaces that are now buried are designated as subcrops. Buried erosion surfaces are termed unconformities, hence subcrops denote the rocks situated right beneath an unconformity. By removing all the overlying rocks above an unconformity, we can create a map of the subcropping units, resulting in a paleogeologic map. Essentially, it represents a geological map as a prehistoric geologist would have documented it prior to the onset of recurrent deposition that obscured the ancient erosion surface. In the cross section of Figure 8, a paleogeologic map depicting the distribution of rocks before the deposition of units XZ would illustrate the eroded exposure of the folded sequence P-T. Figure 9 illustrates the subcropping units depicted in a "greyed out" manner beneath the unconformably overlaying sequence. The intersection of a plane denoting the erosion surface and a geological surface beneath the unconformity constitutes the subcrop limit of the older surface. The intersections between matching structural features on the unconformity and the older surface can be readily identified. The subcrop limit delineates the boundary between an area where the ancient surface is retained beneath the unconformity and an area where it has been eroded. In the first region, the contours of the unconformity structure are elevated relative to the older surface; in the second region, the contours of the unconformity indicate a lower elevation. Numerous hydrocarbon traps exist along the subcrop boundaries of reservoir rocks beneath angular unconformities in the Western Canada Sedimentary Basin. Lighter oil and gas ascend in the updip direction via subsurface reservoirs until they reach impermeable younger strata above an unconformity. The pursuit of such an occurrence is referred to as a subcrop play in the petroleum sector.

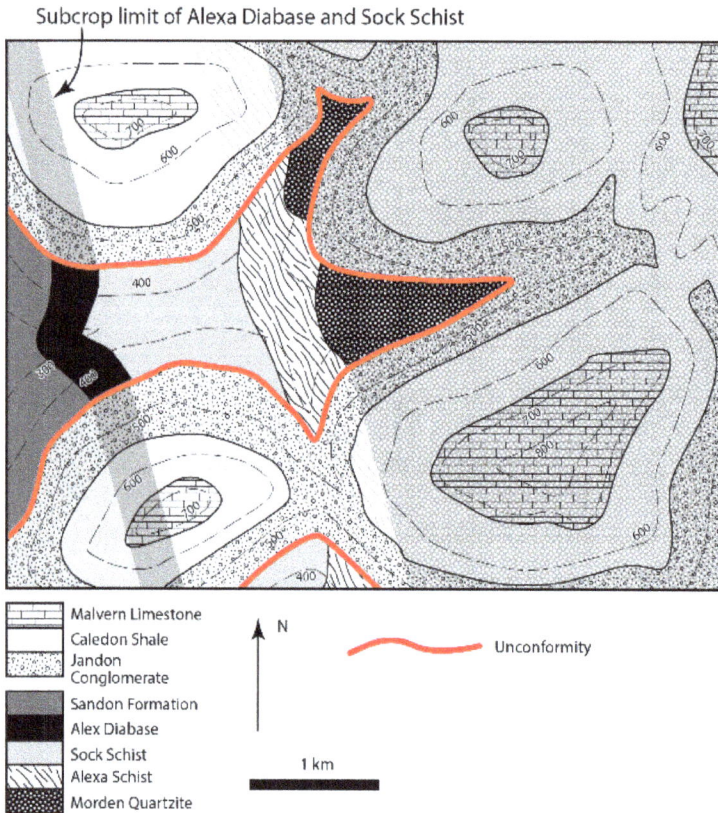

Figure 9. Same geological map as in Fig. 7, but with units above the unconformity (red) shown partially transparent, revealing the subcrops of older units below.

PRIMARY STRUCTURES IN SEDIMENTARY ROCKS

Stratification

Outcrop-scale: bedding, lamination

Strata are often seen at outcrop scale too, and are one of the main characteristics that allow us to recognize sedimentary rocks.

Layers thicker than 1 cm are known as **beds**. A layer thinner than this is a **lamina** (plural **laminae**).

Several sedimentary processes (storms, turbidity currents) suspend large amounts of sediment in a sudden event, and allow it to settle out slowly, producing a **graded bed** with a sharp base, above which the grain-size progressively fines upward. Graded beds are useful indicators of **younging direction**: i.e. in highly deformed rocks, they indicate whether or not the strata are overturned. However, graded beds must be used with caution because there are sedimentary processes that can produce **inverse grading**. For example, densely moving masses of colliding grains, avalanching down a steep slope, tend to interact so that the large grains rise to the top. Therefore graded bedding should only be used as a way-up indicator if it is seen multiple times in a succession of strata.

Structures generated by currents, way-up indicators

We have already mentioned graded bedding as an indicator of way-up, or younging direction. Many other sedimentary structures can be used as way-up indicators.

Bedforms and cross-stratification

Amongst the most useful is **cross-stratification**. Cross-stratification results from the migration of **bedforms** during sedimentation. Bedforms are waves on the bedding surface produced by the action of either currents or waves.

Bedforms are generally classified into larger forms, called **dunes,** and smaller types called **ripples.** By convention:

- cross-stratification produced by dunes is called **cross-bedding**;
- cross-stratification produced by ripples is called **cross-lamination.**

Cross-stratification indicates way-up most effectively because it produces **truncations** of laminae, that resemble small-scale angular unconformities. The laminae that are truncated are always *below* the truncation surface.

Even if truncations cannot be seen, it's sometimes possible to use cross lamination by noting that most bedforms have ridges that are sharper – more pointed – than the troughs. However, in deformed rocks it's sometimes the case that the curvature of surfaces is modified. Truncations therefore give a more certain indication of way-up than lamination shape.

Figure 10. Ripples. Cambrian Gog Group, Lake Louise, Alberta.

Figure 11. Dunes. Recent sediments of Kennetcook River Estuary, Nova Scotia.

Figure 12. Cross-lamination. Carboniferous Horton Group, Tennycape, Nova Scotia.

Figure 13. Cross-bedding. Triassic Fundy Group, Burntcoat Head, Nova Scotia.

Figure 14. Cross-bedding truncation indicates that these strata are upside down. Banff, Alberta.

SOLE MARKINGS

Sole markings are a second type of sedimentary structure that is useful for structural geologists. Sole markings are formed when coarse sediment (usually sand) is deposited rapidly (usually by a current) on a muddy substrate. Sole markings include:

- **Groove casts**: grooves are made by currents dragging objects across the mud; these are then filled by sand and preserved as molds on the base of a sandstone bed;

- **Flute casts**: currents produce scoop-like depressions in the mud that fade out in a down-current direction; these are then filled by sand and preserved as molds on the base of a sandstone bed;

- **Bioturbation structures (trace fossils)**: horizontal burrows and trails are filled by sand and preserved as molds on the base of a sandstone bed.

- **Load structures**: bulges in the bottom of a sandstone bed formed when denser sand sinks into less dense wet mud.

Note: strictly speaking, all these structures should be called molds, not casts. However, the term 'cast' is more commonly used.

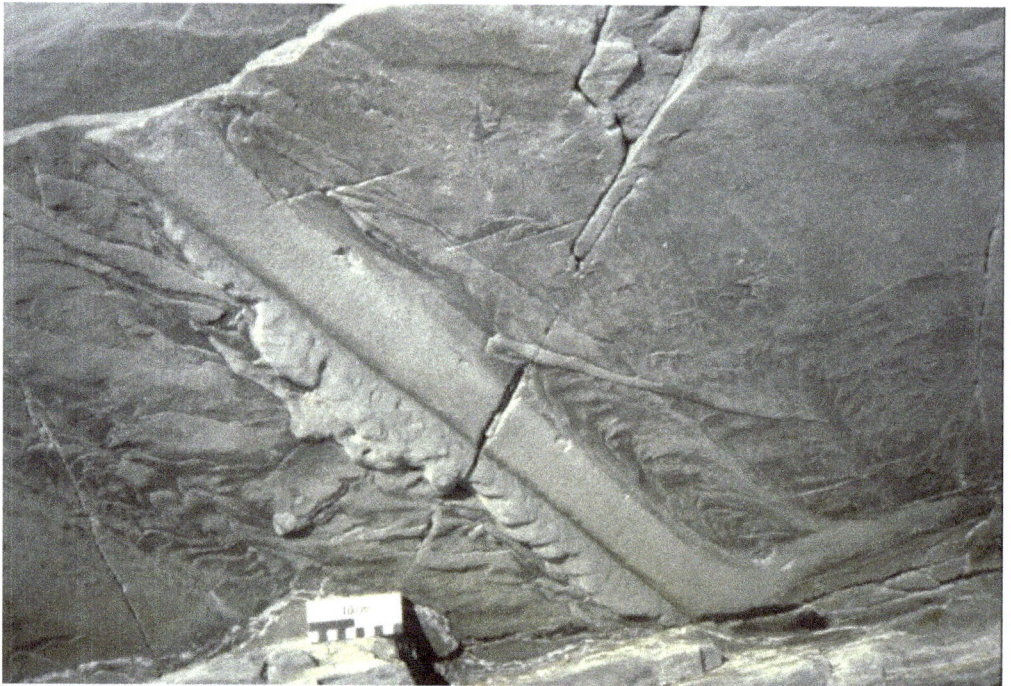

Figure 15. Groove casts. Cambrian Goldenville Group, Nova Scotia.

Figure 16. Flute casts. Cambrian Goldenville Group, Nova Scotia.

Figure 17. Bioturbation structures. Cambrian Goldenville Group, Nova Scotia.

Structures generated by soft-sediment deformation

Recently deposited silt may undergo deformation when in an unlithified state. Soft-sediment deformation structures pose challenges for structural geologists because to their resemblance to tectonic structures, which develop post-lithification of the sediment. Several categories exist.

- **Mudcracks:** these are formed by shrinkage of mud as it dries out. Mudcracks are most visible when they are filled by overlying sediment that is different. They thin downwards to a point and therefore can be good way-up indicators.

- **Load structures:** bulges in the bottom of a sandstone bed formed when denser sand sinks into less dense wet mud. Load structures also fall into the category of **sole markings.** Corresponding narrow tongues of mud penetrating upward between load structures are called **flame structures.**

- **Convolute lamination:** sand that is deposited under water is often initially very loosely packed. Subsequently, the grains may settle into a denser packing, and the water between the grains escapes upward. As this occurs, the water may **liquidize** the sand. Any lamination may become deformed into complex chaotic folds. Convolute lamination is only a good way-up indicator if it is truncated by younger laminae.

Figure 18. Mudcracks (top view). Ordovician Aguathuna Formation, Port au Port Peninsula, Newfoundland.

- **Slump structures:** sediments that are deposited on a slope may undergo catastrophic slope failure, and start to move under the influence of gravity. Beds may become tightly folded as a result of this type of process. Slump folds can be difficult to distinguish from tectonic folds, and are not particularly effective as way-up indicators. However, if the way-up is known from other structures, slump folds can be used in sedimentology and basin analysis as indicators of **paleoslope.**

Figure 19. Mudcracks (side view). Carboniferous Mabou Group, Lismore, Nova Scotia.

Figure 20. Load structures. Ordovician Lower Head Formation, Newfoundland.

Figure 21. Convolute lamination. Ordovician Eagle Island Sandstone, Newfoundland.

Figure 22. Slump folds. Isparta province, Turkey.

PRIMARY STRUCTURES IN IGNEOUS ROCKS

Intrusions

Intrusions by their nature do not typically show stratification, and cannot usually be used to determine tilting or way-up. However, in unravelling the structural history of a complex region, it is important to know the relative timing of intrusions, and this is where contact relationships are all-important.

Exocontact features are formed in the **host rock** (also known as **country rock)** by the effect of an intrusion. A metamorphic **aureole** (baked zone) is often recognizable from changes in texture or mineralogy. There may be minor intrusions where magma has filled cracks branching off the main intrusion. These are called **dykes** (dikes in the US) unless they are parallel to strata in the host rock, in which case they are **sills**.

Endocontact features are formed within an intrusion, where it comes in contact with the host rock. A **chill zone** is typically finer-grained than the bulk of the intrusion. **Xenoliths** are pieces of host rock that broke off and are surrounded by the intrusion.

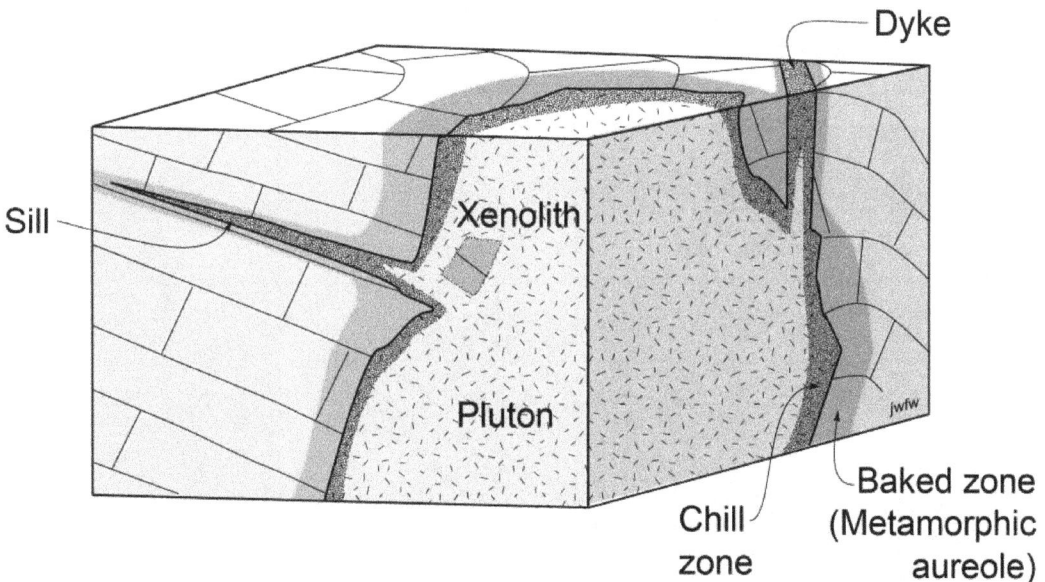

Figure 23. Block diagram showing features of igneous intrusive contacts.

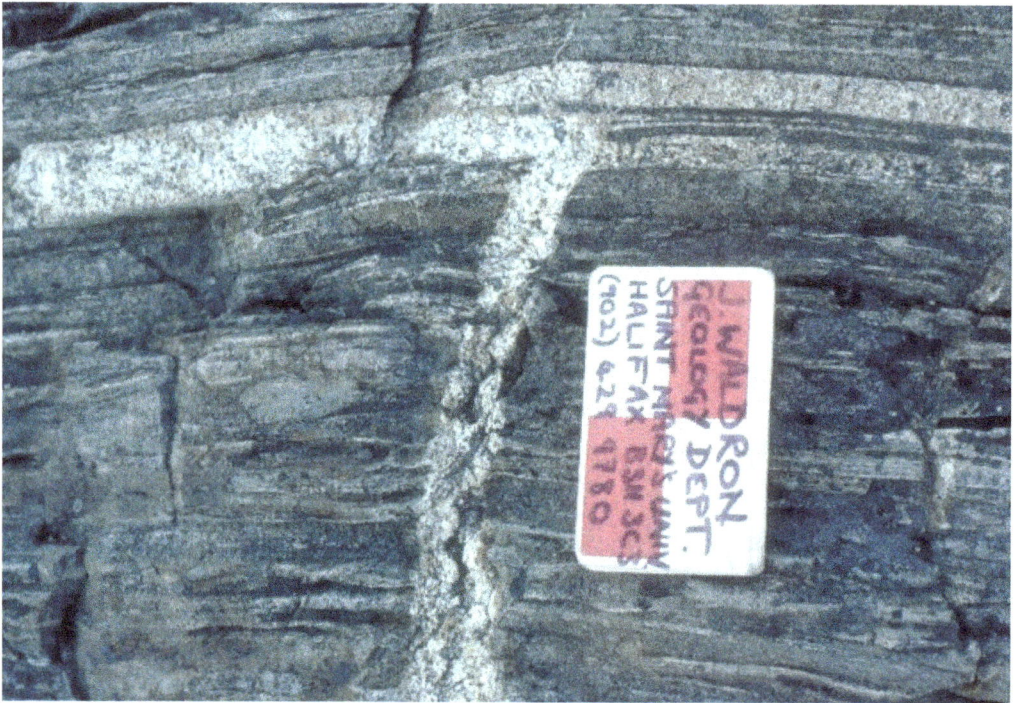

Figure 24. Dyke and sill of Devonian granite intruded into Cambrian Halifax Group, Nova Scotia.

Figure 25. Xenoliths of Cambrian Halifax Group in Devonian Granite, Portuguese Cove, Nova Scotia.

Volcanic rocks

Volcanic rocks are typically stratified, but bedding is often much less clear than in sedimentary rocks. Sometimes the contacts between volcanic flows are conspicuous because they are weathered. Soil layers called 'bole', consisting of soft, clay-rich weathered lava are sometimes visible.

Lava erupted under water typically forms balloon-like **pillows** typically 0.5 – 2 m in diameter, formed by rapid chilling. Later pillows conform in shape to those beneath them in a flow, giving a general indication of way-up.

Thick lava flows may shrink and crack as they cool, producing **columnar joints**. These typically form perpendicular to the base and top of a flow. As a result, the columns are elongated in the direction of the *pole to stratification* and therefore can be used to estimate the orientation of strata where bedding cannot be observed directly.

Note that columnar joints are common in sills too. Sills can be distinguished from flows only by looking at their contacts: sills show intrusive contacts top and bottom, whereas flows typically show one weathered surface.

Figure 26. Pillow lavas. Cretaceous Troodos Massif, Cyprus.

Figure 27. Columnar joints in sill. Salisbury Crags, Edinburgh, Scotland.

LAB 2. CROSS-SECTIONS AND THREE-POINT PROBLEMS

Topographic profiles and cross-sections

Topographic profiles show the shape of the Earth's surface in a view that simulates a vertical slice through the landscape. Topographic profiles may be constructed by noting where *topographic contours* cross the line of the profile.

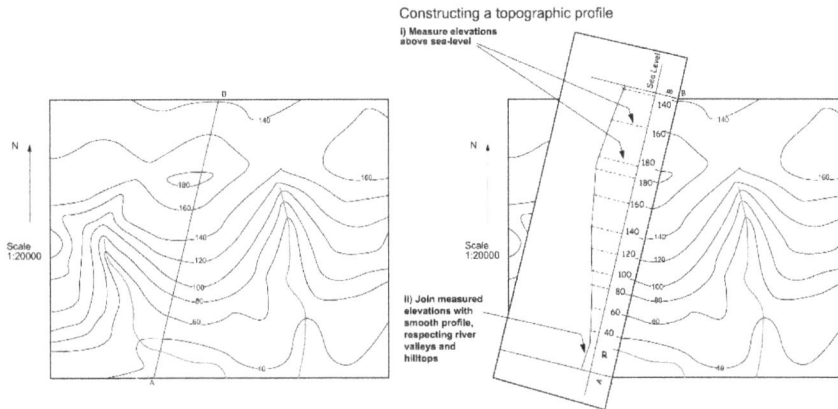

Figure 1. Topographic map, showing technique for drawing a topographic profile along line AB.

You may remember the technique for drawing a topographic profile from your introductory geology course (Fig. 1). On a profile or a cross-section, the ratio of the vertical scale to the horizontal scale, expressed as a fraction, is the **vertical exaggeration**. If the vertical and horizontal scales are equal the section is said to have a **natural scale**. Unless there is a good reason to use vertical exaggeration, *it is generally best in structural geology to draw sections at natural scale.*

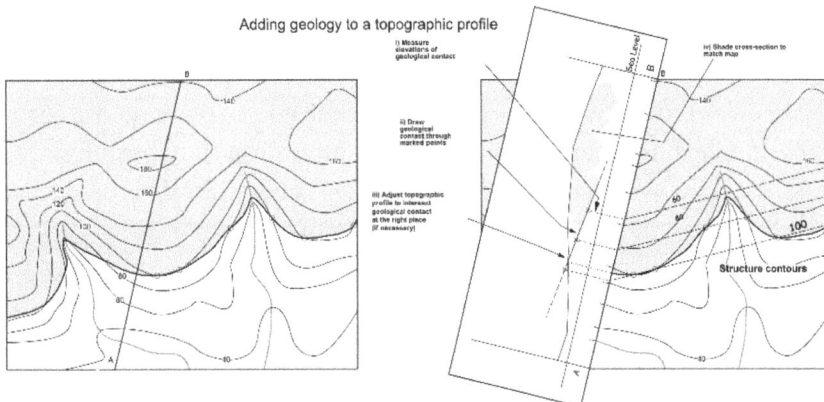

Figure 2. Geologic map, showing technique for adding geology, using structure contours, to make a geologic cross-section.

A vertical crosssection depicting the trace of a geological surface can be built by identifying the intersections of structure contours with the line of section.When a natural scale is employed and the section line is perpendicular to the strike, the crosssection accurately represents the genuine dip.In portions oblique to the strike, the crosssection reveals the apparent dip.It may be demonstrated that the perceived dip is consistently less than the genuine dip.Figure 3 illustrates the apparent dip and true dip across several cross-sections.

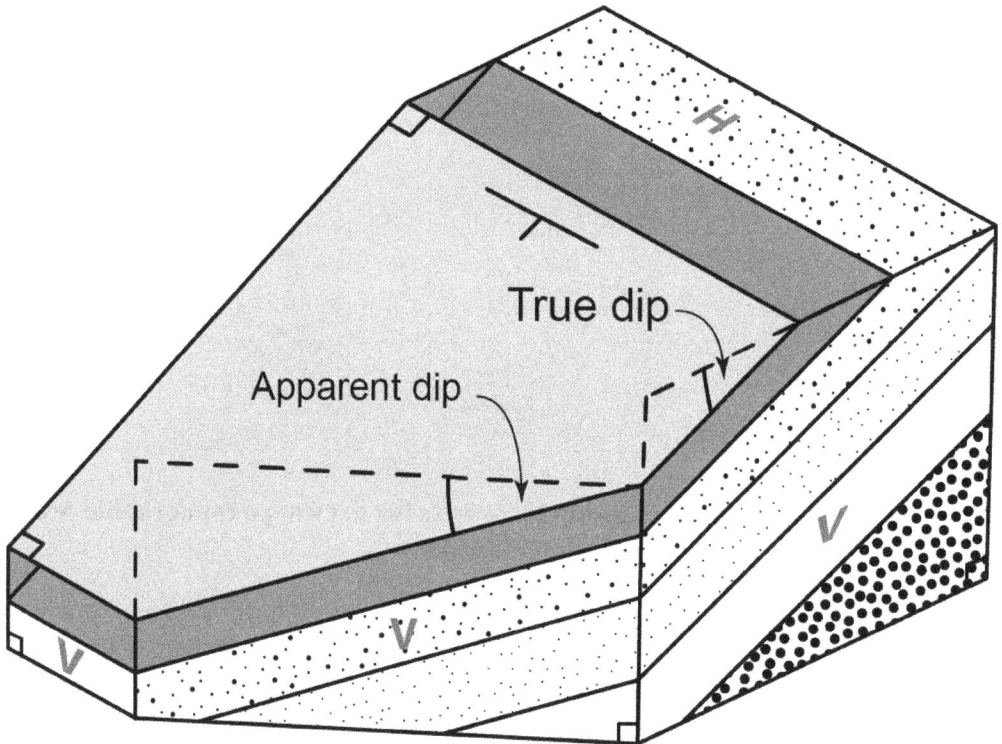

Figure 3. Block diagram illustrating the difference between the true (t) and apparent (a) dip of the stippled plane. Planes labelled H and V are horizontal and vertical respectively, and right angles are labelled in the usual way.

In a cross-section through a succession containing an angular unconformity, there is no possible orientation of the cross-section that will show true dip for both the underlying and overlying parts of the succession, unless they both have the same strike. Cross-sections with vertical exaggeration show neither true nor apparent dip.

Three-point problems

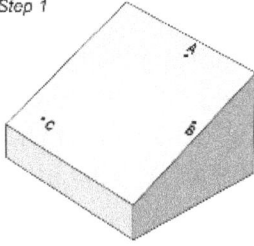

Step 1 — Three points of different elevation

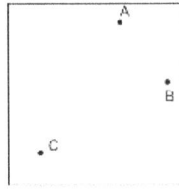

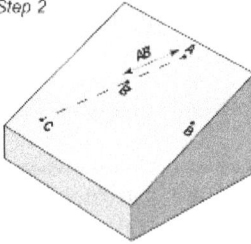

Step 2 — Find the elevation of the intermediate point

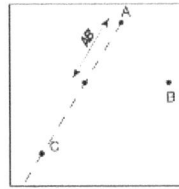

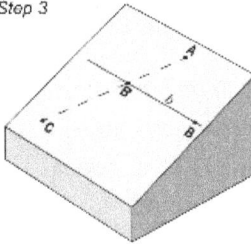

Step 3 — Draw the first contour with intermediate point elevation

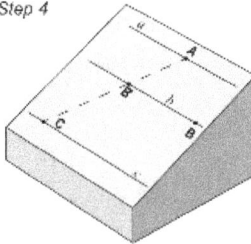

Step 4 — Draw contours parallel to first contour

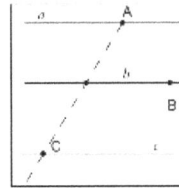

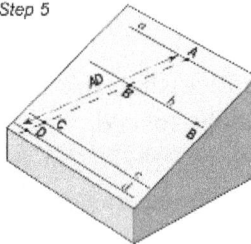

Step 5 — Find the "round number" contours

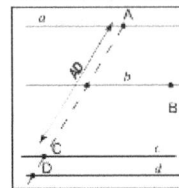

Figure 4. Map and block diagram illustrating solution of three-point problems. A, B, and C are three points at different elevation on the surface. 3-D view on the left, map view on the right.

Structure contours may be drawn for a planar surface if we know its elevation at three points. This is known as a 'three point problem'.

You need three points of known location and elevation all plotted on a map (Fig. 4). One of these points must be the highest – it's labelled A in the diagram. One is the lowest, labelled C. The intermediate point is B. The elevations of the three points are represented in the following by lowercase letters $a, b, c,$ from highest to lowest.

The first step in solving the problem is to connect the high and low points on the map with a line AC. Somewhere along this line there will be a point (call it B') with the same elevation as B. The distance of B' along line AC is proportional to the height differences.

In other words: *(Length AB') / (Length AC) = (a-b) / (a-c)* so *(Length AB') = (Length AC) × (a-b) / (a-c)*

Use this relationship to locate B', and join B and B' with a line. This is your first structure contour.

Since we are assuming this is a planar surface, we can also draw two more contours at elevations a and c

It's unlikely that a, *b or c* is a 'round number' like 200 or 5000, comparable with the topographic contours on the map. There are several ways to find a contour with a round number value (call it d). Probably the easiest is to repeat the calculation above, to locate a point with elevation d that lies on line AC:

(Length AD) = (Length AC) × (a-d) / (a-c)

Note that in the example, point D lies beyond the end of line AC, but this is not necessarily the case; the same method can be used to find contours that pass between A and C.

Assignment

1. * Examine the geological map of the Grand Canyon. Even without structure contours, we can make some inferences about the orientations of different geological units.

    Link to larger version of map

    Look at the topographic contours and notice that their spacing varies dramatically. In some places they are widely spaced, whereas in others they are so close together that they merge together. The steepest slopes are typically found on particular geologic units of erosion-resistant rocks, known as 'cliff-forming' units.

    a) Using the legend, identify and name one cliff-forming Paleozoic unit that outcrops in consistently steep topographic slopes.

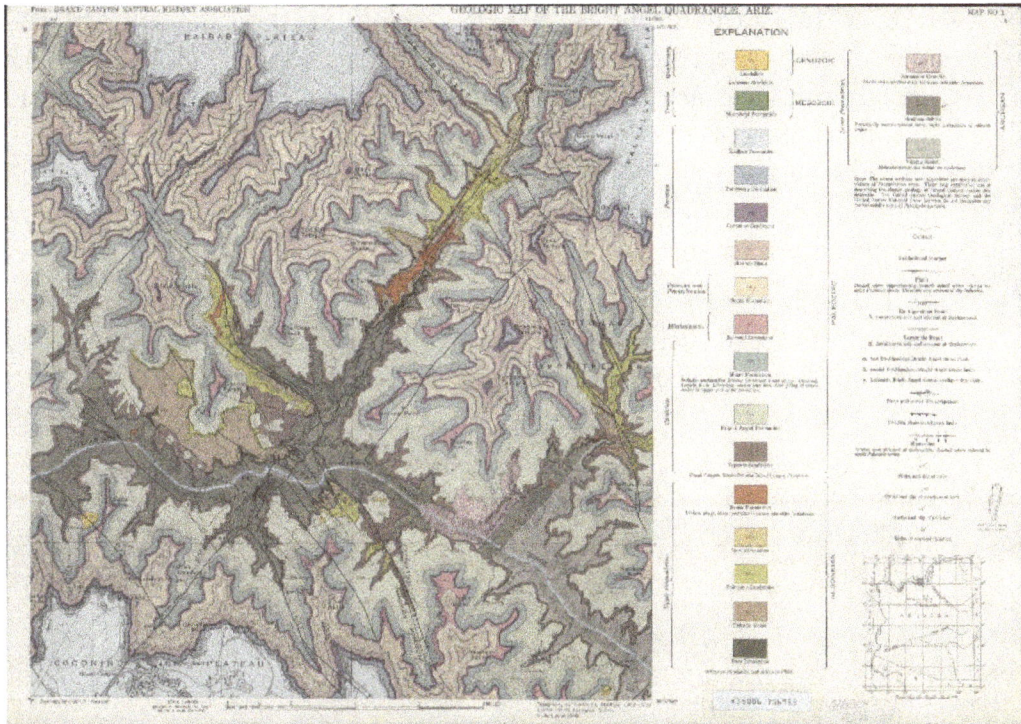

Lab 2 Question 1. Geological Map of the Grand Canyon (Maxson 1961, USGS, 1:48000). Credit: U.S. Geological Survey Department of the Interior/USGS U.S. Geological Survey/Published by the Grand Canyon Association. A version of this map can also be downloaded from https://ngmdb.usgs.gov/Prodesc/ proddesc_33640.htm

In addition to information about erosion-resistance, the map pattern carries information about the dip of units. Based on the map pattern, what can you say about the dip of the following units? (In each case, your answer should be something like 'approximately horizontal', 'approximately vertical', 'dipping gently', etc.)

b) The Archean units

c) The Algonquian units

d) The Paleozoic units

e) The Bright Angel Fault

In addition, the map gives you information about geologic time, both by the principle of superposition (younger rocks on top of older) and by the principle of cross-cutting relationships (older structures are cut by younger). An angular unconformity is a type of cross-cutting relationship where a younger unit lies on the eroded surfaces of many different older units.

f) Look for unconformities that are visible in the map pattern and identify two. In each case, specify which unit lies immediately above the unconformity surface. (This is the best way to specify the location of an unconformity in a stratigraphic succession because typically a single upper unit lies on a variety of lower units). For each unconformity, say which units of rock are overstepped, and also mention any evidence for onlap at the unconformity surface.

g) There is at least one more unconformity on the map but it is a disconformity, so there is no cross-cutting relationship. Using the legend and your knowledge of the geologic timescale, identify its location in the stratigraphy.

2. Map 1 shows the trace of an unconformable contact between slate and an overlying conglomerate. Conformably overlying the conglomerate is sandstone and limestone.

Map 1.

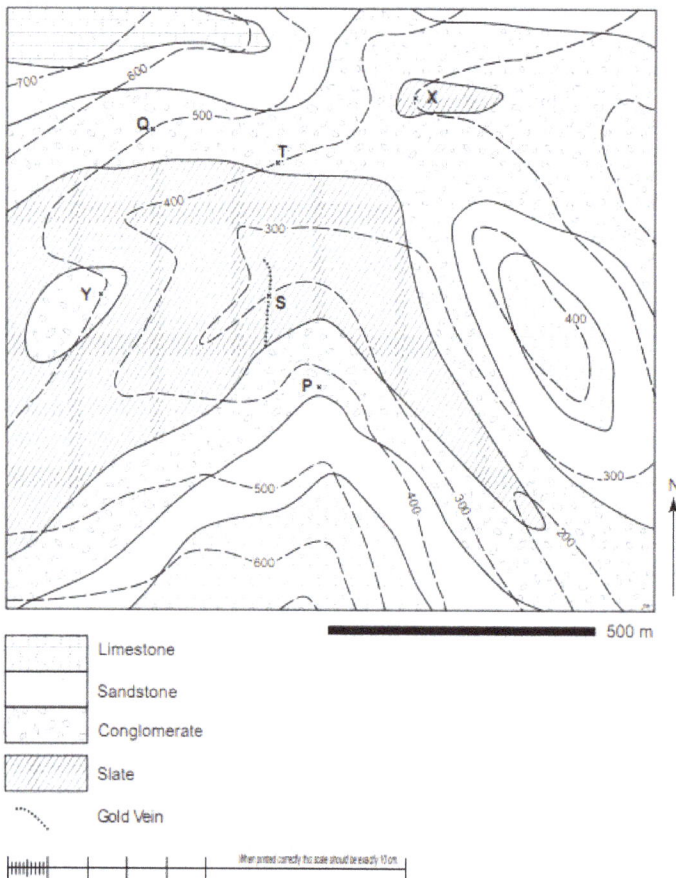

Lab 2, Map 1: Stratified units overlying slate with veins

a. Draw structure contours on the unconformity surface.

b. Determine its orientation (strike and dip).

c. Draw structure contours on the remaining contacts. You may notice that some structure contours are shared between multiple surfaces. Draw the structure contours in pencil and label each surface with a different colour.

d. Draw two vertical topographic profiles with bearings of 099° and 000° through the point P. You may remember the technique for drawing a topographic profile from your introductory geology course. The scale of the map is 1:7500. Your topographic profiles should be drawn at **natural scale** (no vertical exaggeration).

e. Now add the unconformity to the topographic profile to make a cross section. To do this, use the intersections of structure contours with the profile in exactly the same way you used the intersections of topographic contours in the previous question! (Do not try to use the calculated dip to place the plane on the cross section; if the cross-section is at an angle to the dip it will show an apparent dip, not a true dip. By far the easiest and most accurate way to place surfaces on the section is by using the structure contours. Also, the contour technique always works even if you have to construct a vertically exaggerated section.) If you do not have enough contours to constrain the surface on the cross-section, interpolate contours at intermediate elevations (325, 350, 375 m etc.).

f. Complete the sections by adding in the remaining surfaces and shade the units with appropriate patterns.

*g. Which of the slopes of the traces of the unconformity on the above cross-sections equals the true dip and which an apparent dip?

*h. A copy of Map 1 is provided in next week's lab. Enter your answer from part b in the space provided above the map, as you will need to use these numbers.

3. Also on Map 1 is a dotted line representing the trace of a gold-bearing vein in slates exposed on the hillside at S. A planar gold-bearing vein was also inter-sected in borehole Q, 100 m below the topographic surface, and in borehole T, 300 m below the surface. Assuming all three observations are of the same gold vein, you have enough information to determine its orientation.

Use the 3-point method to draw structure contours on the vein.

a. Determine its strike and dip.

b. Draw the subcrop line of the vein by finding the intersection of the two sets of structure contours.

c. Determine the trend and plunge of this line.

d. Complete the outcrop pattern of the vein on the map.

e. Add the vein to both cross-sections.

f. * A prospector suggests drilling through the outlier at Y to look for the gold vein below the unconformity. Explain why this suggestion would be a bad idea.

g. * The prospector then suggests drilling through the inlier at X to look for the gold vein below the unconformity. Explain why this suggestion would also be a bad idea.

h. * Enter your answers from parts b and d in the spaces provided above the map in next week's lab, as you will need to use these numbers.

Map 1.

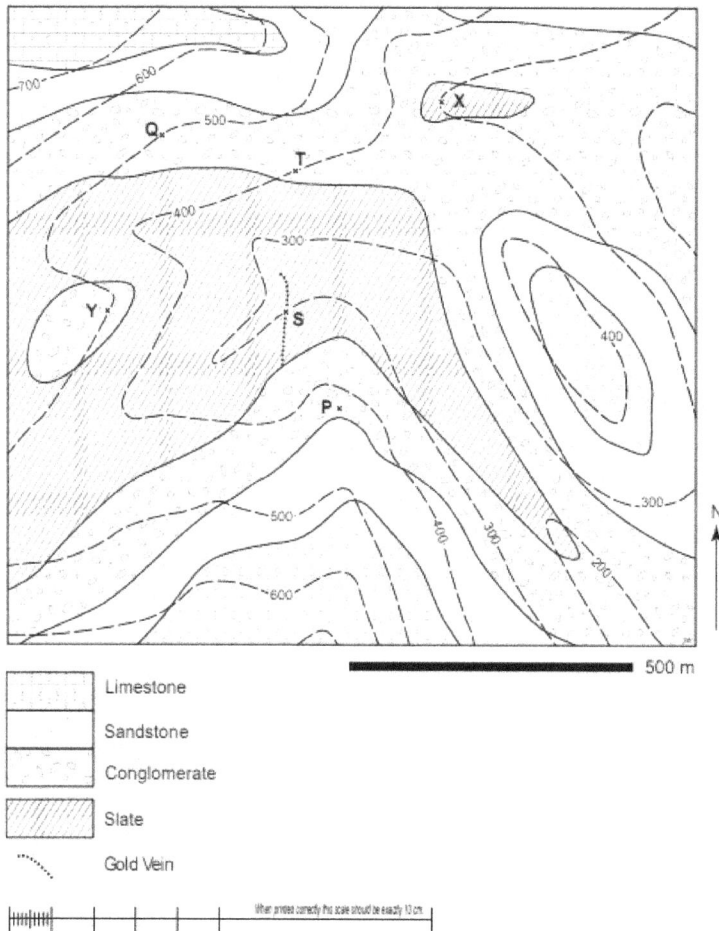

Limestone

Sandstone

Conglomerate

Slate

Gold Vein

Lab 2 Map 1 Copy orientations for Lab 3

4
Chapter

INTRODUCTION TO STEREOGRAPHIC PROJECTION

INTRODUCTION: AN OVERVIEW

Stereographic projection is an effective technique for addressing geometric issues in structural geology. In contrast to structure contouring and other map-based methodologies, it solely maintains the orientation of lines and planes, lacking the capacity to keep positional links. Nonetheless, it is highly beneficial, as orientation issues are prevalent in structural geology. Stereographic projection has been utilized since the second century B.C. and is a prevalent technique employed by crystallographers for depicting crystal morphology. Nonetheless, there exists a significant distinction. Crystallographers utilize an upper hemisphere projection, while structural geologists consistently employ the lower hemisphere. The lower hemisphere signifies the area under the Earth's surface where the rocks remain uneroded. Nevertheless, if you are already familiar with stereographic projection in mineralogy, the lower hemisphere may require some adjustment. Envision gazing into a bowl-shaped concavity on the Earth's surface. The online visualization tool available at https://app.visiblegeology.com/stereonet.html may prove beneficial.

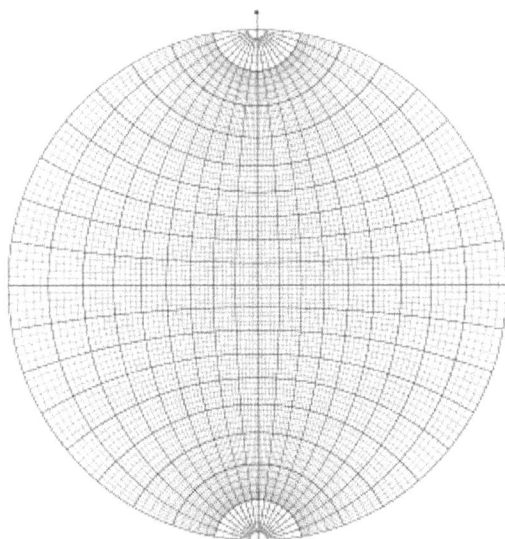

Figure 1. Wullf net for plotting and measuring features on a stereographic projection

STEREOGRAM BASICS

A stereographic projection comprises two components. The projection, or stereogram, is often rendered on tracing paper and depicts a bowl-shaped surface integrated into the Earth. The stereographic net, or stereonet, serves as the three-dimensional counterpart to a protractor. It is utilized to quantify angles on the projection. To measure angles, it is necessary to rotate the net in relation to the tracing paper. For practical purposes, we typically rotate the tracing paper while keeping the net stationary; however, it is crucial to acknowledge that, in actuality, the projection maintains a fixed orientation, necessitating the rotation of the net for accurate measurements. To create a stereogram, utilize a sheet of tracing paper and delineate a circle that matches the radius of an existing stereonet. This circle is referred to as the primordial. Indicate the center with a cross and place a north arrow at the top of the page on the primitive. Mark points E, S, and W (or 090, 180, and 270 degrees) in 90° intervals around the primitive. Reinforcing the center with adhesive tape on the reverse side of the tracing paper might be beneficial at times. The stereonet can be fortified with cardstock to prolong its durability. It is practical to insert a traditional thumbtack into the center of the net. Safeguard yourself and others against the thumbtack by securing it within an eraser while not in use. Multiple versions of stereonets exist. We will commence with a Wulff net, utilized for the creation of the correct, or equal-angle, stereographic projection. Subsequent laboratories will employ a Schmidt net, which creates an equal-area projection.

Principle of stereographic projection

For stereographic projection, a line or a plane is imagined to be surrounded by a **projection sphere** (Fig. 1a). A plane intersects the sphere in a trace that is a **great circle** that bisects the sphere precisely. A line intersects the sphere in a point. To image features on a sheet of paper, these traces and points are projected from a point at the summit or **zenith** of the sphere onto the equatorial plane. This is clearer in a diagram (Fig. 2b), which shows the method for stereographic projection of a dipping plane. A family of planes dipping at various increments is shown in Fig. 3a. Planes project as curves that are actually perfectly circular arcs called **cyclographic traces** or just **great circles.** Lines project as **points** also known as **poles.**

As a general principle, planes that dip at low angles are represented by great circles having significant curvature and lying closer to the primitive, whereas steeply dipping planes are characterized by straighter great circles passing close to the centre of the plot. All vertical planes will project as straight lines passing through the centre of the stereogram.

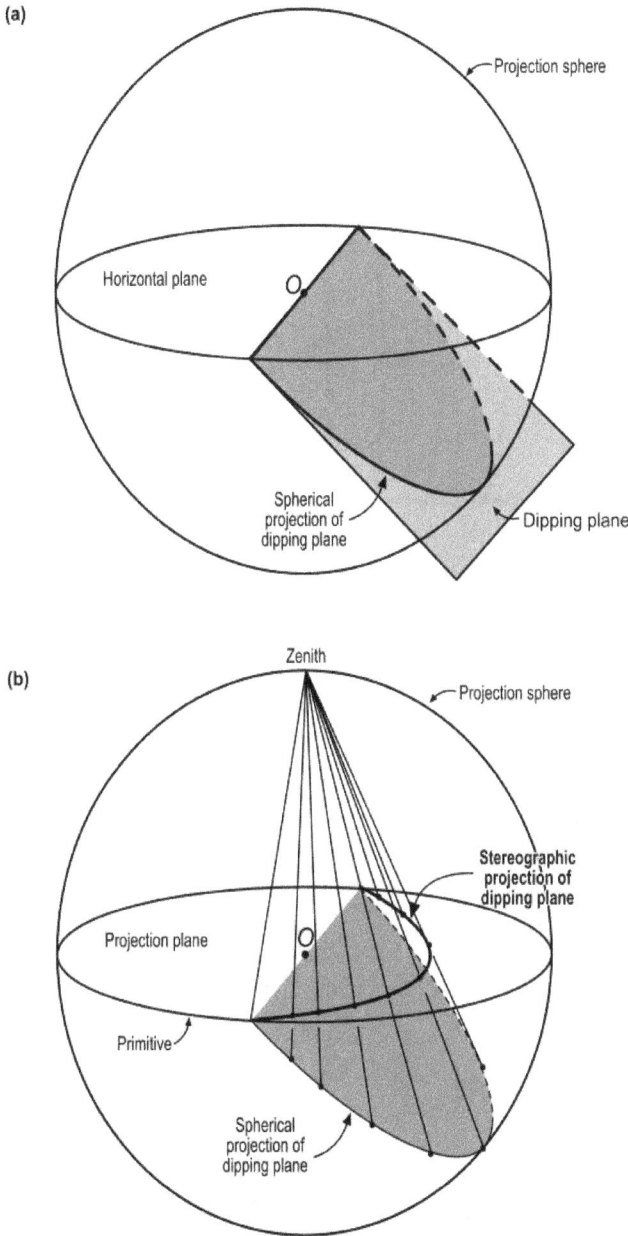

Figure 2. Principle of the stereographic projection.

Sometimes we represent a plane by its **pole**. The pole to a plane is the plot of a line perpendicular to the plane. For a horizontal plane the pole is in the centre of the net. Gently dipping planes have poles near the centre; steeply dipping planes have poles near the edge. *The pole is always in the opposite quadrant to the great circle.* Poles are used when plotting numerous great circles would make the plot cluttered and confusing.

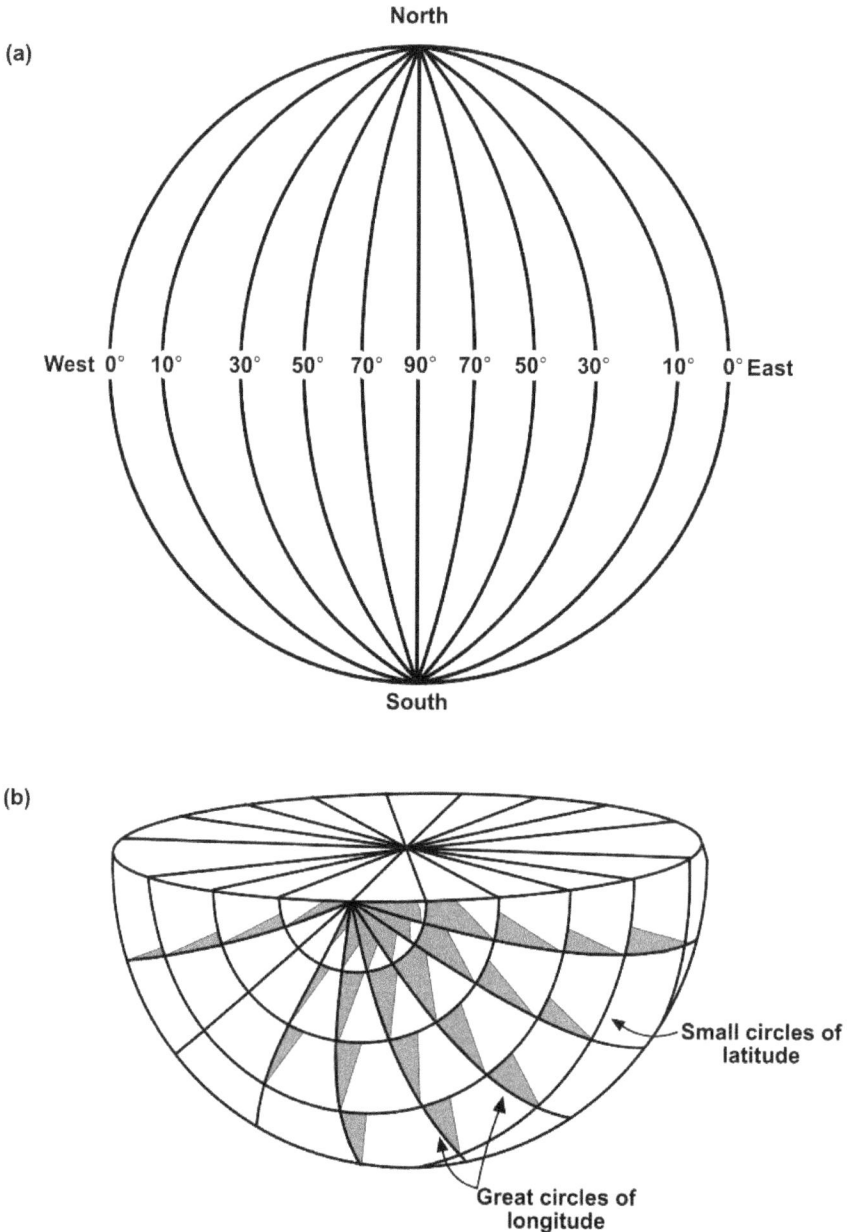

Figure 3. Great and small circles.

FEATURES OF THE NET

The stereographic net assists in the construction of great circles and points. It contains a family of great circles intersecting at the top and bottom points of the net (Fig. 3a). Every fifth cyclographic trace is bolder so that ten degree increments can be easily counted. By rotating the net, a great circle can be maneuvered into any desired strike and dip orientation. The net also contains **small circles** (Fig.

3b) that are helpful in solving rotation problems, and act as a scale of pitch angles along each great circle.

One great circle on the net corresponds to a vertical plane and is straight. It runs from top to bottom of the net. One trace in the family of small circles is also straight, from left to right on the net; it is actually a great circle too. These two intersecting lines form four **straight radii** which are crucial for counting angles of dip and plunge.

When the plot is located so that its north arrow coincides with the top of the net, it is said to be in the **reference position.**

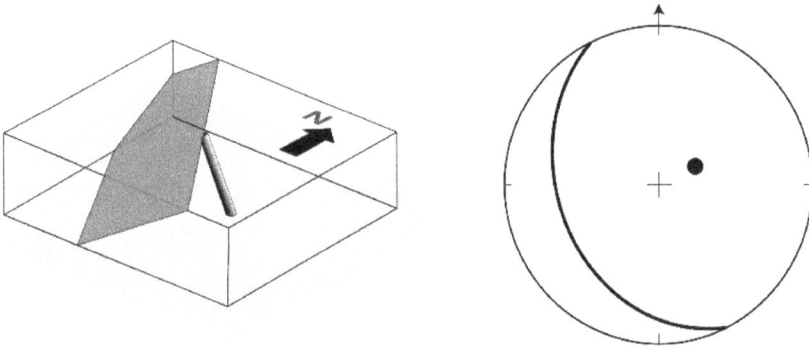

Figure 4. Block diagram and stereographic projection of a plane and its pole.

TWO APPLICATIONS OF THE STEREOGRAPHIC PROJECTION

True and apparent dip

In lab 2 you encountered the terms 'true dip' and 'apparent dip'. Refer back to lab 2 if you need to remind yourself of the difference. In principle, it's possible to make conversions between true and apparent dip by trigonometry. However, it's generally much easier to make the conversion using the stereographic projection. The construction is shown in Fig. 5.

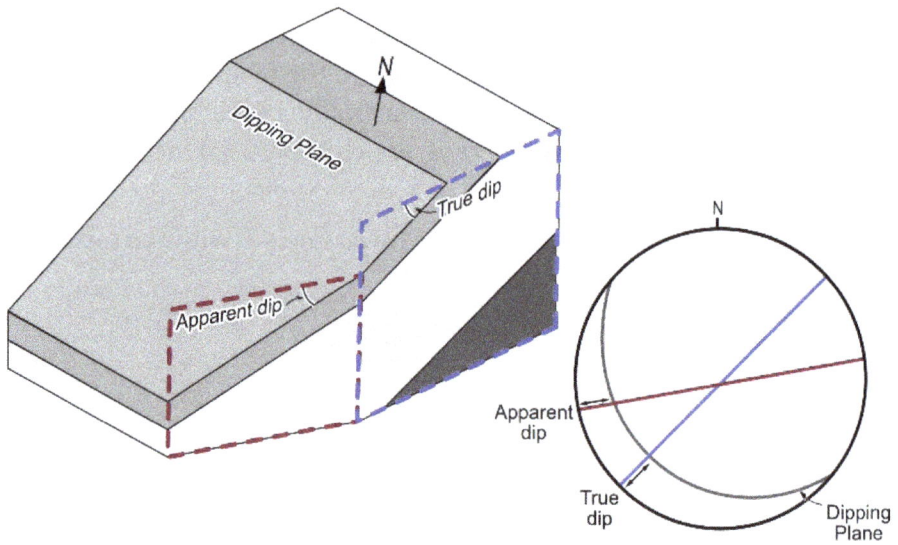

Figure 5. Converting between true and apparent dip with the stereographic projection.

True and apparent thickness

A prevalent issue in stratigraphy is ascertaining the actual thickness of a formation. True thickness is quantified at a right angle to the bedding plane. Frequently, in the field or within a subsurface well, an apparent thickness is measured—one that is oblique to bedding, hence overestimating the true thickness (Fig. 6). Numerous trigonometric techniques exist for determining true thickness, contingent upon the specific measurement conditions.. However, one method using the stereographic projection works every time: multiply the apparent thickness by the cosine of the angle (θ) between the pole to bedding and the line of measurement (the trend and plunge of the traverse, tape measure, or whatever measuring device was used).

True thickness = Measured thickness × cos(θ)

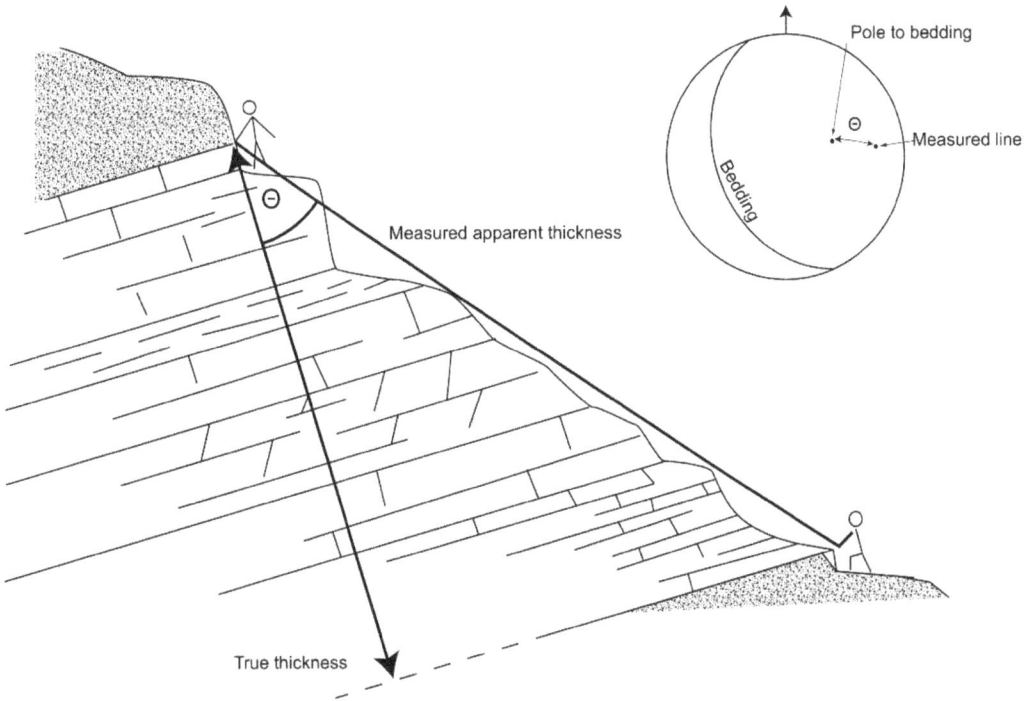

Figure 6. Calculating the true thickness of a carbonate unit, using measured apparent thickness and the pole to bedding. True thickness = Measured thickness × cos(θ).

LAB 3. WORKING WITH STEREOGRAPHIC PROJECTIONS

Basic plotting operations

Setting up the projection

You will need a Wullf net.

For use in this course, your net must be 15 cm in diameter. If you download your own, print the Wullf net on a plain sheet of paper at at 100% scale. (*Many computers will automatically shrink the image to fit a smaller paper area by default. Use the options in the printer dialog on your computer to print at 100%.*)

Next, use an an old fashioned thumb tack to make a small hole through the exact centre of the net. Then insert the tack pointing upwards, through the hole.

Now, take a sheet of tracing paper, and press it over the net so that the thumb tack makes a neat hole in the centre.

Then, trace the circular outline, the edge of the net, known as the primitive, on the tracing paper. Mark the four cardinal points N, E, S, W, on the tracing paper[1] with small 'ticks', and distinguish N with an arrow.

You should now be able to rotate your projection (on the tracing paper) over the net, and then return it to the *starting position* with N at the top.

Plotting the orientation of a line

The simplest geometric information one can display on a stereographic projection is the orientation of a vertical line. It projects as a point in the middle of the stereonet. The next easiest information to portray is the orientation of a line that has a trend due north or due south. With the overlay in the reference position, count the amount of plunge from either the north or south index mark (as appropriate) on the primitive towards the center of the net along the N-S line, and place a dot on the tracing paper at this position. Remember, a line is represented by a point on the stereogram.

The procedure for a line of general orientation is as follows:

a) Visualize the problem first, using a pencil. Imagine that the pencil originates at the centre of the net, and extends downward to intersect a "bowl" below the net. In which quadrant will it intersect the hemisphere, and whether it will be close to the primitive or far away?

b) With the overlay in the reference position, make a mark on the primitive that corresponds to the trend of the line.

c) Rotate the overlay until the mark is aligned with a straight radius of the net; count the angle of plunge inwards from the primitive along the straight radius, and make a small cross.

d) Return the overlay to the reference position and check that the cross is in the expected general position.

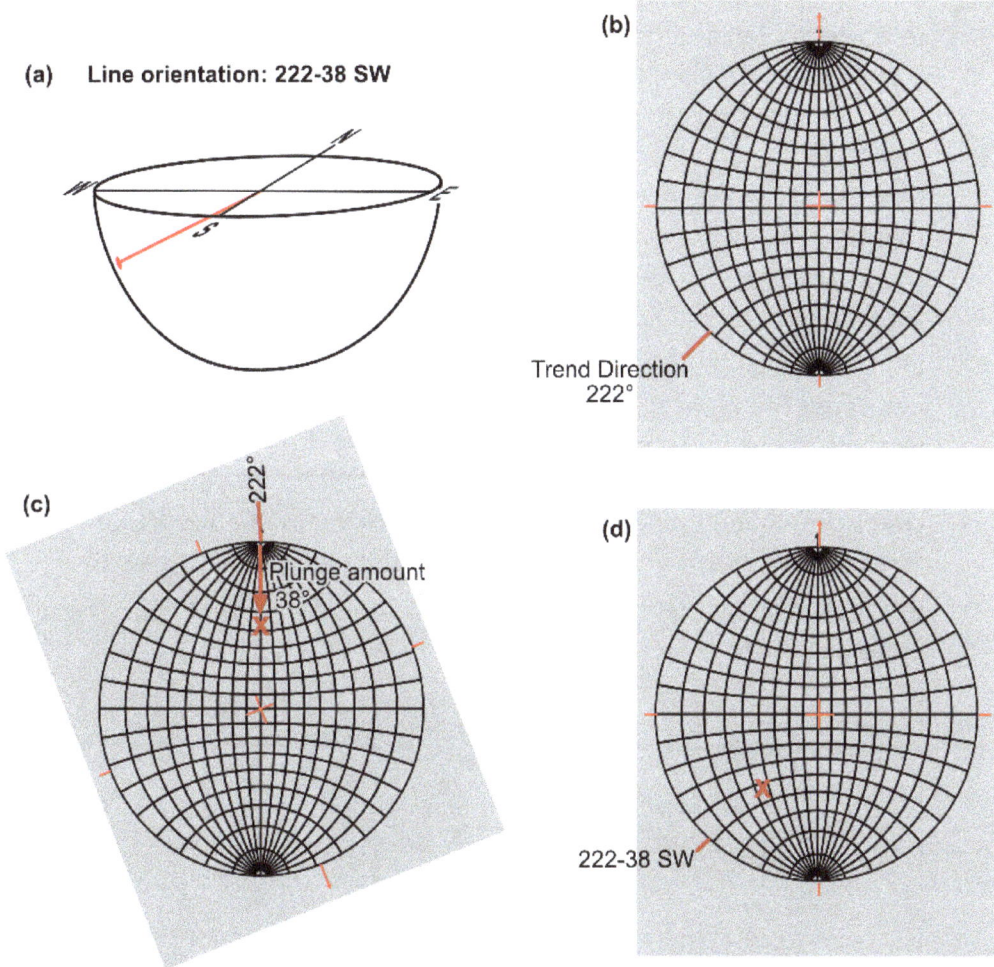

(a)   Line orientation: 222-38 SW

(b)

Trend Direction
222°

(c)

222°

Plunge amount
38°

(d)

222-38 SW

Figure 1. Procedure for plotting the stereographic projection of a plunging line. Plunge of 38 and trend of 222. Tracing paper shown grey.

Example problem: Plot the point L representing the line 300-50. On your projection label the angles corresponding to the trend and plunge.

Plotting a plane and its pole

Horizontal and vertical planes are straightforward to plot. A horizontal plane is represented by the primitive. N-S and E-W striking vertical planes are represented by the straight lines on the stereonet connecting the N-S poles and E-W positions, respectively. Other vertical planes are straight diameters oriented parallel to the strike.

To plot an inclined plane follow these steps:

> a) Visualize the problem using your hand or a piece of paper. Imagine that this plane object passes through the centre of the plot and intersects

a hemisphere below the net. In which quadrants will the curved line of intersection lie, and how close to the primitive will that curve be? Where will its pole be? The pole should be on the opposite side of the net. If the great circle is near the primitive then the pole will be near the centre, and vice versa.

b) With the overlay in the reference position make a mark on the primitive that corresponds to the strike direction of the plane. As a matter of sound practice, you should always make this mark on the strike direction according to the right hand rule.

c) Rotate the stereonet until the mark is aligned with the top point on the net. **From the right hand side of the net**, count degrees inward along the straight radius, until you reach the amount of dip. Trace the great circle which passes through this point; optionally, to plot the pole to the plane, count the same number of degrees **outward from the centre along the straight radius towards the left side of the net** and mark a small 'x'.

d) Return the overlay to the reference position and check that the great circle corresponds with the anticipated orientation.

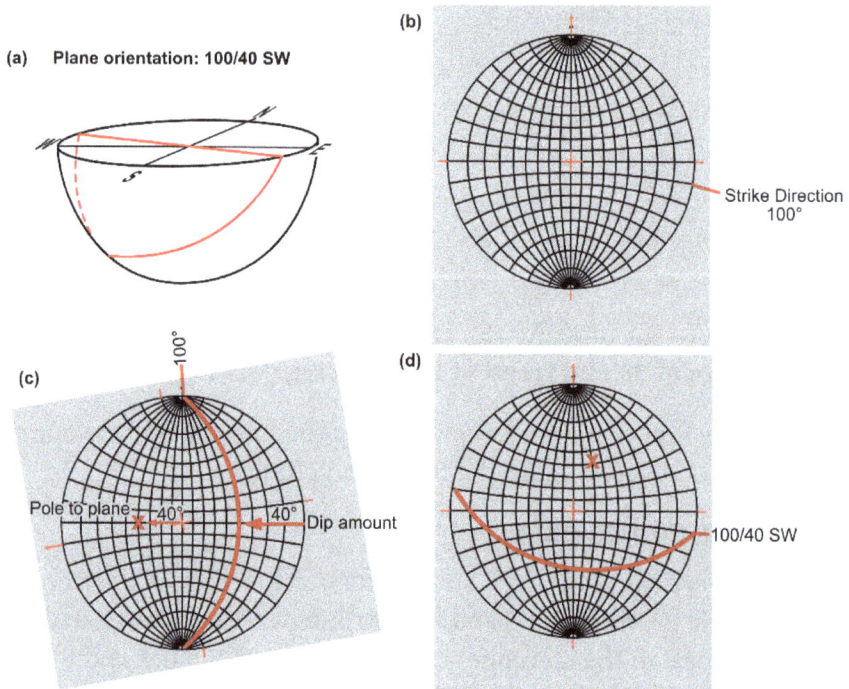

Figure 2. Procedure for plotting the stereographic projection of an inclined plane with a strike of 100, a dip of 40, and the pole to the plane. Tracing paper shown grey.

Example problem: Construct the great circle representing the plane 120/50, and its pole. On the projection label the angles corresponding to the plane's strike, dip and numerical dip direction.

Plotting a line in a plane by rake or pitch

A line that lies in a plane plots as a point on a great circle.

> a) Plot the plane as a great circle following steps a-c of the previous procedure.
>
> b) From the top point of the net, count degrees along the great circle you have just traced until you reach the desired angle of rake.

Measuring the orientation of a line from its pole

Finding the orientation of a line from its pole is just the reverse of the plotting procedure above.

> a) Rotate the net to place the pole on any of the four straight radii. Mark a point on the primitive at the end of this straight radius.
>
> b) Count degrees of plunge inward from the primitive until you come to the pole. Record the plunge.
>
> c) Return the net to the reference position. Count degrees clockwise round the primitive until you come to the mark made in step a. This is the trend.

Example problem: measure the orientation of the line you plotted in the first example and see if you get the answer you started with!

Measuring the strike and dip of a plane from its great circle

Measuring the strike and dip represented by a great circle is just the reverse of the plotting procedure for a plane.

> a) Rotate the net until the plane is on a great circle on the right hand half of the net. (This ensures an answer consistent with right-hand rule.)
>
> b) Mark the end of the great circle that is at the top of the net.
>
> c) Count degrees of dip inward from the right hand side along the straight radius until you come to the great circle.
>
> d) Return to reference position and note the azimuth of the mark made in step b. This is the strike.

Measuring the strike and dip of a plane from its pole

A pole is perpendicular to the plane it represents, so steps a and c are done in a way that seems opposite to the procedure when working with a great circle, or trace!

a) Rotate the net until the pole is on the straight radius on the *left side* of the net.

b) Mark the end of the great circle that is at the top of the net.

c) Count degrees of dip *outward* from the centre along the straight radius toward the *left* hand side of the net until you come to the pole.

d) Return to reference position and note the azimuth of the mark made in b. This is the strike.

Example problem: measure the orientation of the plane you plotted in the first example by working backwards from its pole and see if you get the answer you started with!

Constructions on the stereographic projection

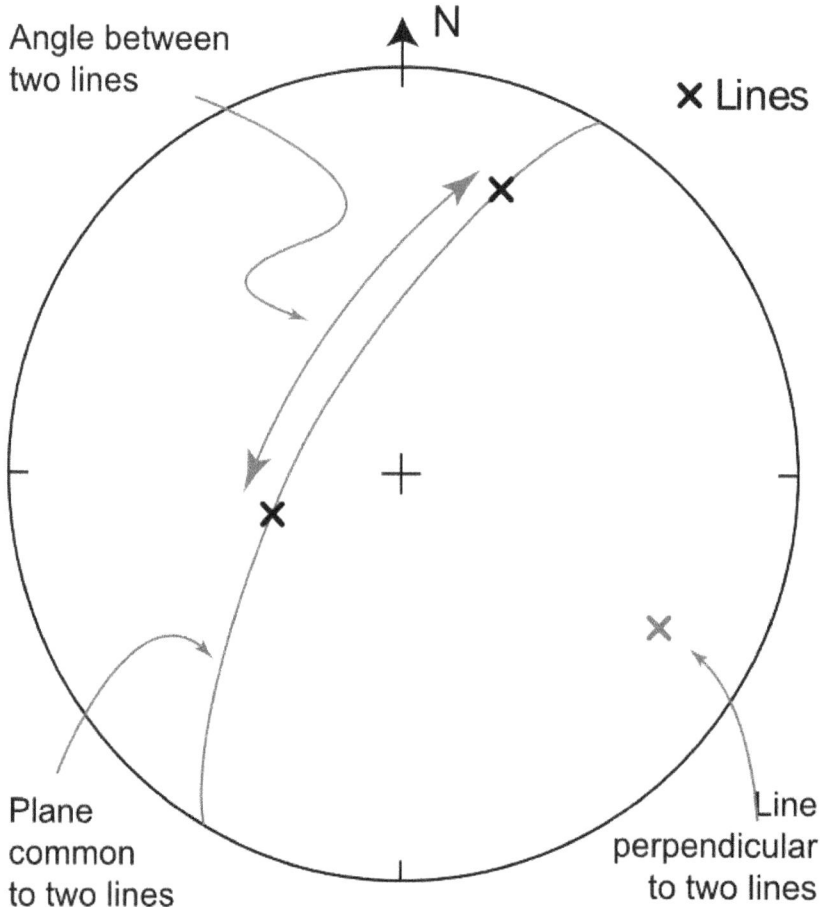

Figure 3. Calculations involving lines.

Finding the plane common to two lines

If you have two lines plotted as points on the stereographic projection, the plane common to the two lines plots as a great circle that passes through both points.

a) Plot both lines as points.

b) Rotate the net so that both poles lie on a single great circle.

c) Trace the great circle representing the plane.

Finding the line perpendicular to two other lines

For any two differently oriented lines, there will be a third line that is perpendicular to both of them.

a-c) Repeat steps a-c above.

d) Find the pole to the plane; it represents the perpendicular line.

Angle between two lines

It's possible to measure the angle between the two lines by counting 2-degree squares along the great circle that passes through both.

a-b) Repeat steps a-b above.

c) Count degrees along the great circle between the two lines.

Note that, unless the lines are at 90°, there will always be *two* answers: one greater than 90° and one less than 90°.

Example problem: Construct the plane common to the lines 318-34 and 206-78; determine the angle between them; also determine the strike and dip of the common plane.

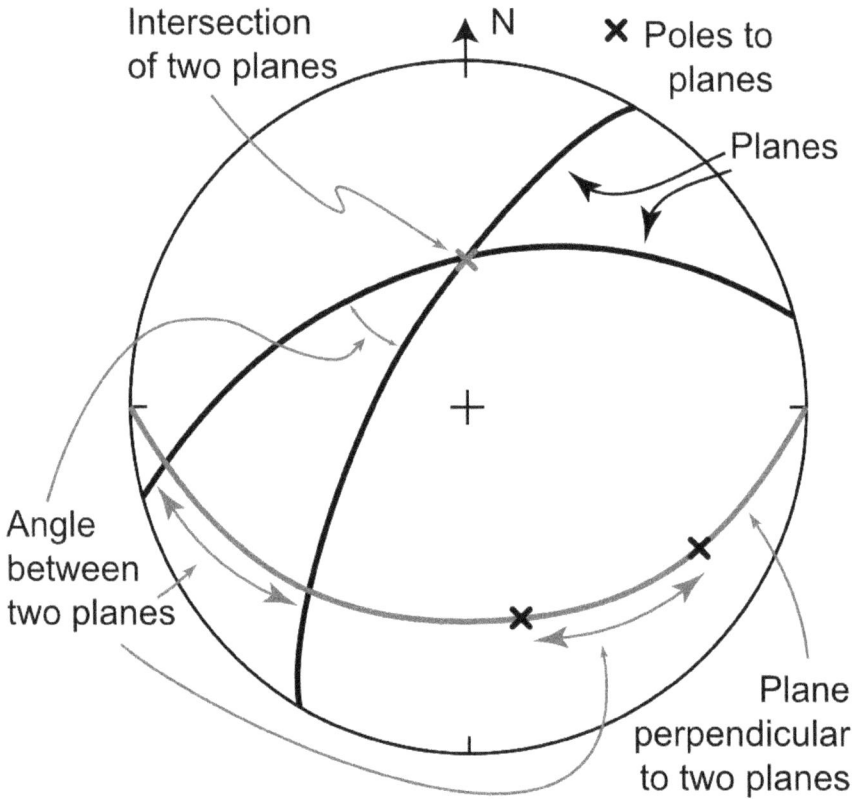

Figure 4. Calculations involving planes.

Line of intersection of two planes

The line of intersection of two planes is represented by the point where their great circles cross.

a) Plot the two intersecting planes as great circles on the overlay.

b) The great circles intersect at a point on the overlay, which represents a line of intersection of the two planes. Determine the orientation of the line.

Plane perpendicular to two planes

To find the plane perpendicular to two other planes, we first find their line of intersection, and then use it as the pole to a third plane.

a) Repeat step a above

b) Repeat step b above, but instead of determining the trend and plunge, move the point of intersection to the straight radius on the left side of the net.

c) Count the number of degrees outward from the centre of the net, along the left straight radius to this point.

d) Count the same number of degrees inward from the primitive along the right straight radius, and trace the great circle that passes through this point.

Angle between two planes

Multiple techniques exist to determine the angle between two planes.While it is feasible to measure the angle between the two great circles using a protractor, this method lacks precision.Two approaches are advised.The initial way is more straightforward to conceptualize, whereas the subsequent method is more expedient. Select whatever option you favor.

First method

a-d) Find the great circle perpendicular to the two planes as above.

e) Locate the point where each original plane intersects the new great circle.

f) Count the number of degrees along the new great circle between these points.

Second method

a) Plot both planes as poles.

b) Rotate the net so that both poles lie on a single great circle.

c) Count degrees between the two poles.

Important note: there are always two possible answers to the angle between two planes. The two angles will add up to 180°. The only way to figure out which one is the right answer to a given geological problem is to visualize the problem in 3D!

Example problem: Plot the line of intersection of the planes 132/22 and 074/68. Also plot a third plane perpendicular to both, and find the angle between the two planes. Determine the orientation of the line of intersection of the first two planes. Determine the orientation of the perpendicular plane.

Answers to the example problems

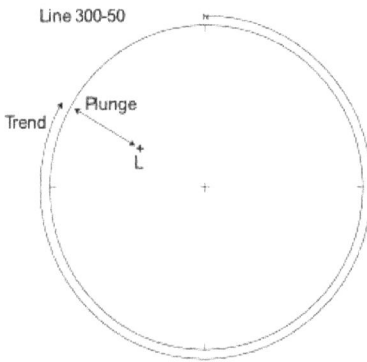

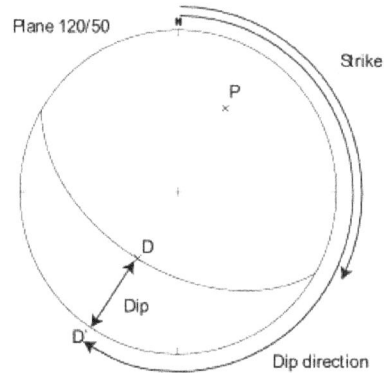

Line 300-50

Trend / Plunge
L

Plane 120/50
Strike
P
D
Dip
D
Dip direction

Intersection of the planes 132/22 and 074/68

L
58°

Intersection line 245-20
Perpendicular plane 335/70

Plane common to the the lines 318-34 and 206-78

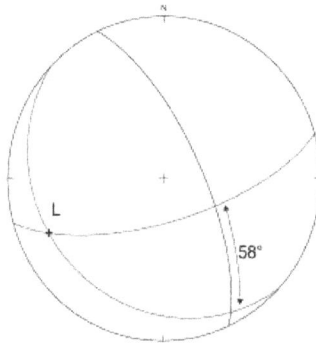

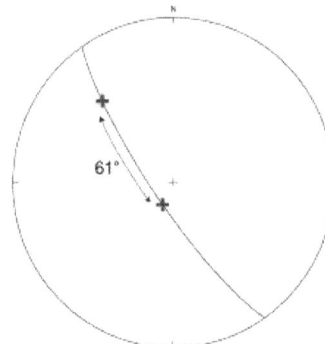

61°

Common plane 145/79

The above plots represent the answers to the following problems:

1. Plot the point L representing the line 300-50. On your projection label the angles corresponding to the trend and plunge.
2. Construct the great circle representing the plane 120/50, and its pole. On the projection label the angles corresponding to the plane's strike, dip and numerical dip direction.
3. Plot the line of intersection of the planes 132/22 and 074/68. Also plot a third plane perpendicular to both, and find the angle between the two planes. Determine the orientation of the line of intersection of the first two planes. Determine the orientation of the perpendicular plane.
4. Construct the plane common to the lines 318-34 and 206-78; determine the angle between them; also determine the strike and dip of the common plane.

Assignment

1. *Plot and label the position of the following lineations on a stereographic projection:

   a) 020-50, b) 295-10, c) 110-80, d) 170-00, e) 210-90.

2. *On a separate sheet of tracing paper plot and label the position of the following inclined planes and their poles:

   a) 310/40, b) 025/85, c) 134/04, d) 265/00, e) 130/90.

3. *A slab of rock is set up in the lab. It shows a lineation and a foliation.

   (a) Measure the strike and dip of the foliation, and the plunge and trend of the lineation with a compass/clinometer.

   (b) Test your accuracy by plotting both on a stereographic projection. First, plot the foliation as a great circle on a stereographic projection. Next, plot the plunge and trend of the lineation as a point. If your lineation lies on the great circle exactly, your measurement is excellent! More likely, it will fall a little off, reflecting the difficulty of measuring with total precision.

   c) Measure the rake of the lineation with a protractor. Now plot the rake measurement, as a point that lies exactly on the great circle. Measure the angle between the two measurements of the lineation. This is an indication of the precision of your measurements.

   d) Use the result to rate your compass use. (Be honest – this is about assessing your own precision correctly, not the initial measurement!)

Error	Rating
<2°	Master of the clinometer
2-5°	Good
5°-10°	Satisfactory
10°-20°	Could use more practice
> 20°	Probably best to try again

4. Along a vertical railroad cutting, a bed shows an apparent dip of 20° in a direction 298°. On level ground outside the cutting a geologist can measure the strike of the beds as 067, but cannot tell which way they are dipping. Use stereographic projection to determine the true, right-hand-rule strike and dip of the bed.

5. On Ashman Ridge in British Columbia, two geologists measured the distance from the bottom to the top of the Quock Formation as 205 m. The measuring tape had a trend of 187° and a plunge of 20°. If the strike and dip of the Quock Formation was 247/63 N, calculate the true thickness of the formation.

The map you worked on last week contains a gold vein cut by an unconformity. Determine the orientation of the subcrop of the vein, by stereographic projection.

(a) Plot the orientation of the gold vein as a great circle. Plot the unconformity as a second great circle. These values (from last week's lab) will be given to you by your instructors.

(b) Mark the point where the two great circles intersect. This represents the orientation of the subcrop line. Determine the trend and plunge of this line.

(c)* Does it agree with the answer you obtained last week by contouring?

(d) Determine the angle between the vein and the unconformity.

Data from previous lab

Map 1.

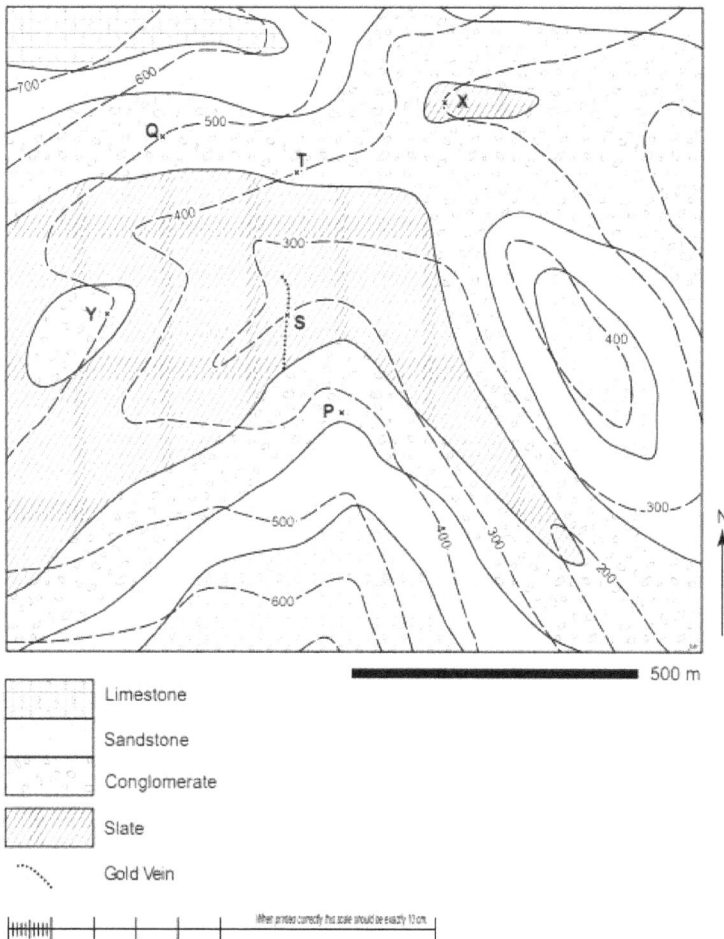

500 m

	Limestone
	Sandstone
	Conglomerate
	Slate
	Gold Vein

When printed correctly this scale should be exactly 13 cm.

Copy of Lab 2 Map 1

Orientation of unconformity: _____

Orientation of vein:_____

Orientation of subcrop line:_____

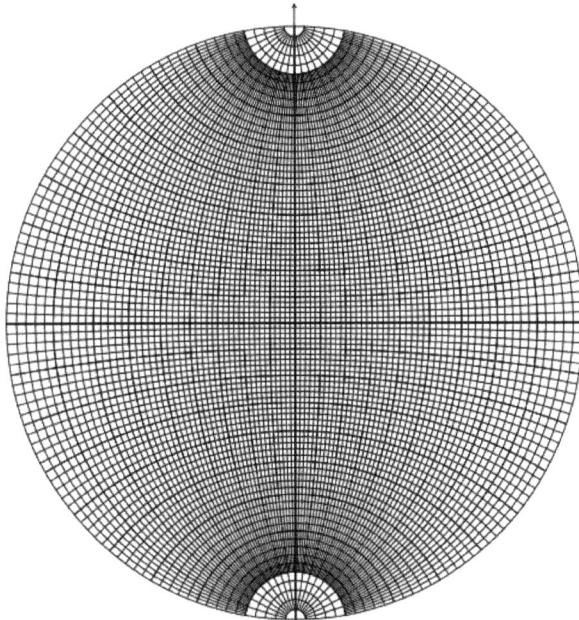

Wulff Net

WULFF NET 15 cm

5
Chapter

INTRODUCTION TO FOLDS

INTRODUCTION: AN OVERVIEW

Folds are among the most remarkable and impressive characteristics of the Earth's crust. Folds typically develop in regions where stratified rocks have been compressed. Folds are prevalent in orogenic belts, areas where the Earth's lithosphere has experienced shortening due to tectonic plate movements. Orogenic belts frequently constitute mountain ranges. Folds hold significant economic importance. Numerous petroleum traps are situated within anticlines. Gold-bearing quartz veins in many significant goldfields were deposited in voids (saddle reefs) that formed between strata during folding. We will prioritize geometry initially, followed by an exploration of kinematics and dynamics once we establish a solid descriptive framework. Folds exhibit considerable variability in style. Consequently, numerous characteristics exist to delineate and quantify in standard folded rocks. Subsequent sections will differentiate between variable and invariant characteristics of folds. Invariant features are unaffected by fold orientation, while variant features are contingent upon the orientation of a fold. Invariant features are emphasized upon their initial introduction in the subsequent sections. All words in the subsequent sections are uniformly relevant to folds at microscopic, outcrop, and map scales.

GEOMETRIC DESCRIPTION OF SINGLE FOLDED SURFACES

The simplest drawing of a folded surface is in **profile** view: an 'end on' view of the fold. Technically, the profile view is a view projected onto a plane perpendicular to the fold hinge.

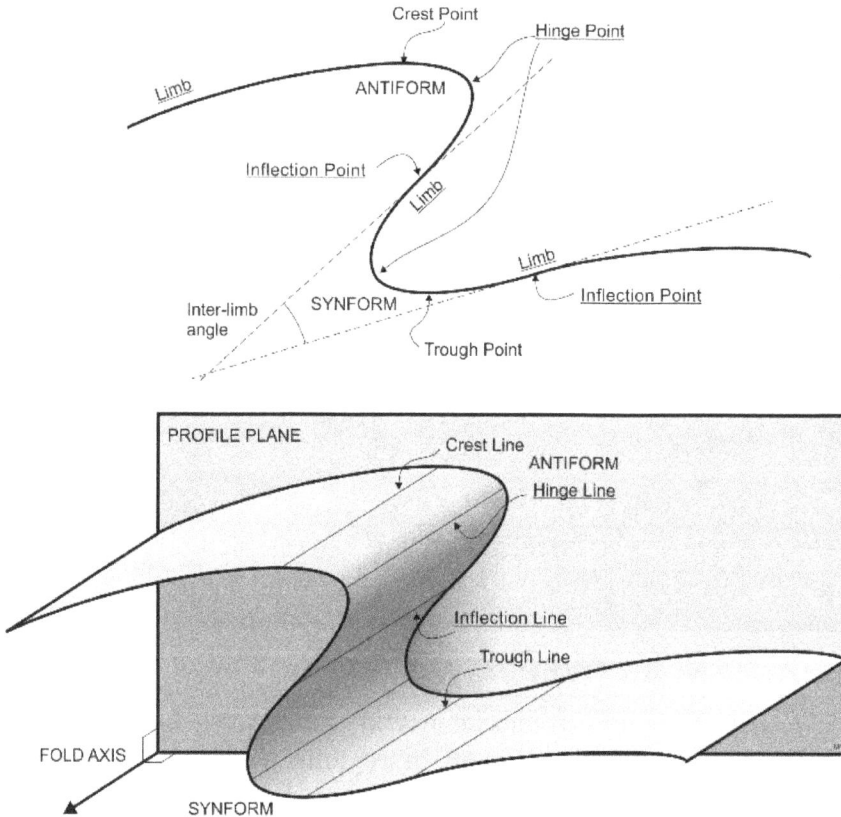

Figure 1. Features of single folded surfaces in 2 and 3 dimensions.

SINGLE FOLDED SURFACES IN PROFILE

Invariant points

Points of tightest curvature are **hinges**, located in the **hinge region.** Points of minimum curvature are **inflection points** which are located on fold **limbs.** If a tangent is drawn at the inflection point on each limb of a fold, the tangents intersect at the **inter-limb angle.**

Inter-limb angle

Folds can be classified by inter-limb angle as:

Category	Inter-limb angle
Gentle	180 – 170°
Open	170 – 90°
Tight	90 – 10°
Isoclinal	10 – 0°
Ptygmatic or Fan	< 0°

Note that there are several different schemes for defining these categories. The above scheme is that given in the text by Davis et al. (2011).

FOLD SHAPE

Based on their shape, fold hinges can be classified as **angular** or **rounded**. (There are more sophisticated classifications but these will do for our purposes.)

Closing direction: synforms and antiforms

If the limbs dip away from the hinge, then the fold **closes upward;** we say the fold is an **antiform**. If the limbs dip towards the hinge, then the fold **closes downward**, and the fold is a **synform.**

Notice that if a fold closes 'sideways' so that one limb dips toward the hinge and the other dips away, it's not possible to define it as either an antiform or a synform. (We will meet the term 'recumbent' which is helpful in describing such folds, in a later subsection.)

Also in profile, we can identify the highest point on the trace of an antiform, called a **crest point**. The lowest point on the trace of a synform is called a **trough point**. Note that the crest and trough points do not necessarily coincide with the hinge points, except in perfectly angular, or perfectly upright folds.

Younging direction: synclines and anticlines

A folded surface in sedimentary rocks has a stratigraphic top side and a stratigraphic bottom side. These define the **younging direction**. If the younging direction is *towards the inside* of the fold, then the fold is a **syncline.** If the younging direction is *away from the inside* of the fold, then the fold is an **anticline.**

In areas of mild deformation like the Rocky Mountain foothills, where the rocks are regionally the right way up, anticlines are antiforms; the terms can be used more or less interchangeably. Similarly, in such areas synclines are also synforms.

However, in areas like the interior of the Cordillera, or the Alps of Europe, there are regions where the rocks are upside down over large areas. There, it's possible for antiforms to be synclines and synforms to be anticlines. Therefore it is important that *if you don't know the younging direction for sure, use only the terms antiform and synform to describe the fold geometry!*

Possible combinations of closing and younging direction:

	Antiform	Synform
Anticline	Antiformal anticline	Synformal anticline
Syncline	Antiformal syncline	Synformal syncline

Single folded surfaces in 3-D

In 3-D all the variant and invariant points can be extended into lines.

Hinge line: a line representing the locus of maximum curvature (i.e. joining all the hinge points.) The plane perpendicular to the fold hinge is called the **profile plane.** It's the plane on which we draw a profile view of a fold.

We also recognize **inflection lines, crest lines** and **trough lines.**

In very regular folds, all these lines are straight and parallel (they have the same trend and plunge). Indeed all planes measured on such folds are parallel to this unique **generating line.** We describe such folds as **cylindrical,** and the direction of the unique line is the **fold axis.**

Not all folds are cylindrical, however. In **non-cylindrical folds** the hinge lines, inflection lines, trough lines and crest lines may be curved and irregular. The highest point on a crest line is described as a **culmination point**; layers dip away from a culmination point in all directions. A culmination point is often an attractive target for petroleum drilling. The lowest point on a trough line is a **depression point**. Layers dip towards the depression point in all directions. If a culmination point is particularly symmetrical, such that the dip increases at about the same rate in all directions, then the fold is a **structural dome.** The opposite of a dome, a fold with a symmetrical trough point, is a **structural basin.** Structural domes and basins are sometimes described as **periclines.**

Fold attitude

All the lines in a fold have a trend and plunge; in cylindrical folds these are all the same – the fold axis orientation. In non-cylindrical folds, the easiest and most regularly measured feature is the hinge line. We can classify folds by hinge-line plunge as follows:

Plunge	Classification by hinge-line plunge
0-10°	Subhorizontal
10-30°	Gently plunging
30-60°	Moderately plunging
60-80°	Steeply plunging
80-90°	Subvertical

Trains of folds in profile and 3-D

The above terms apply to single folds. Commonly, multiple folds are observed in the same folded surface, in which case certain patterns are common.

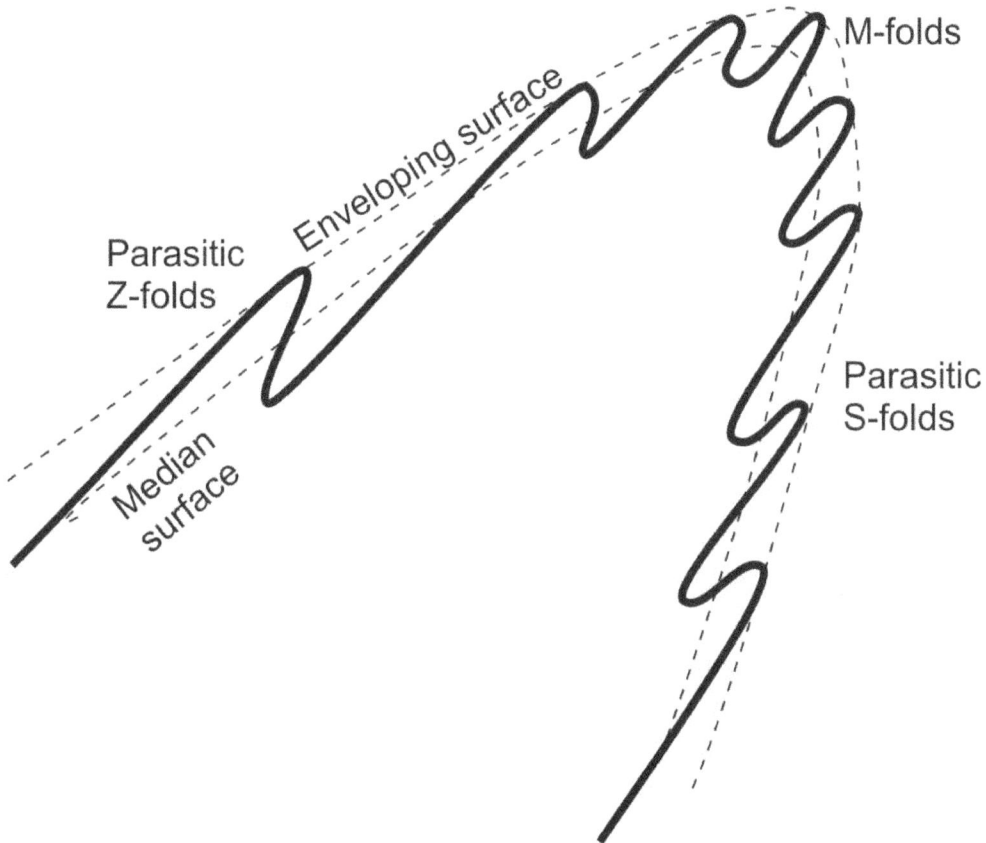

Figure 2. Parasitic folds.

If we draw a surface that is tangent to all the folds, it is called an **enveloping surface**. Similarly, the surface that passes through all the inflection lines is a **median surface**.

Many folded surfaces show alternating long limbs and short limbs, making the folds **asymmetric**. This asymmetry allows folds to be labelled as either S-folds or

Z-folds. To determine whether a fold train has S or Z asymmetry look at three limbs with a long-short-long sequence. If the 3 limbs have the asymmetry of the letter Z then this is a **Z-fold**; if the limbs have the asymmetry of the letter S then the fold is an **S-fold**. *Note that this has nothing to do with the fact that the S is rounded and the Z is angular; it is the rotational symmetry that is important.*

A fold that has S-asymmetry when viewed from one end has Z asymmetry from the other. Therefore it is important that we say which way we are looking when we characterize a fold as S or Z. If the direction of viewing is unspecified, then the convention is to look in the *down-plunge* direction.

One common occurrence of S and Z folds is when folds occur at different scales in the same layer. Small folds developed on the limbs of larger folds, with the same orientation, are called **parasitic folds** (Fig. 2). Typically, parasitic folds on one limb have S asymmetry whereas those on the other limb have Z asymmetry. There is often a region on the hinge where parasitic folds have no clear asymmetry; they are described as **M (or W) folds**. Sometimes mapping the asymmetry of outcrop-scale parasitic folds is a good way to locate the hinges of less obvious map-scale major folds.

Features of successive surfaces

It is rare to see just a single folded surface in outcrop. Most layered rocks have multiple layers that are folded together, which enables us to define some additional variant and invariant features.

In profile

A line drawn through all the hinge points of a fold is called the **axial trace** of the fold. (It's also called the **hinge trace**, which is arguably a better term because the trace has little relation to the fold axis, but most structural geologists use axial trace.)

The other variant and invariant points also correspond to traces on the profile plane: **inflection trace, crest trace, trough trace**. (Because all these lines are the traces, on the profile plane, of surfaces in 3-D, the word 'surface' is sometimes added: inflection surface trace, crest surface trace, trough surface trace.)

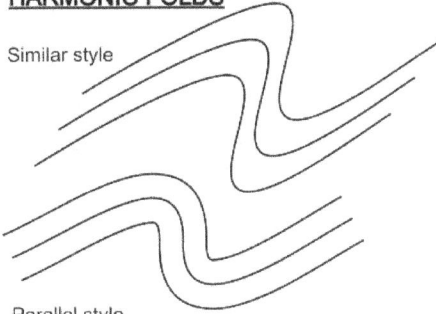

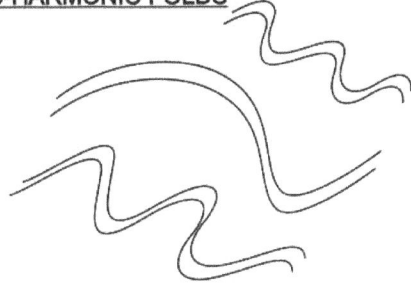

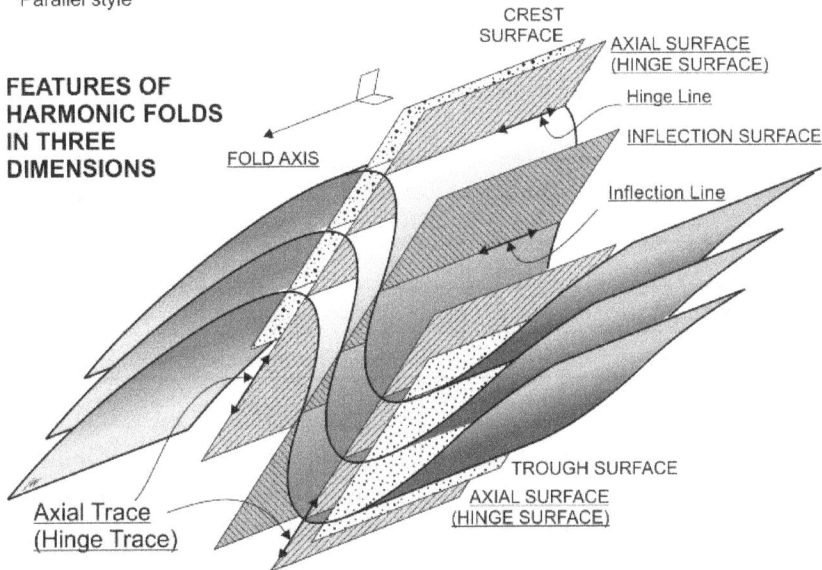

Figure 3. Features of multiple folded surfaces.

IN 3-D

By now you can anticipate that in 3-D each of the lines that we identified on a single surface can be extrapolated into a surface in 3D.

The most important surface is the **axial surface** (or **hinge surface**). This surface is of great kinematic significance because *the pole to the axial surface is often the direction of maximum shortening.*

The other surfaces: **inflection surface, crest surface, and trough surface** are all parallel to the axial surface in a cylindrical fold.

The **facing direction** is the *direction of younging in the fold axial surface.* It is perpendicular to fold hinges but lies in the axial surface. Notice that folds may have overturned limbs but still face upward.

The relationship of facing direction to anticlines, synclines, antiforms, and synforms:

	Antiform	Synform
Anticline	Antiformal anticline	Synformal anticline
	(Upward facing)	(Downward facing)
Syncline	Antiformal syncline	Synformal syncline
	(Downward facing)	(Upward facing)

Notice that in upward facing folds the '-form' and the '-cline' are the same. In downward facing folds the '-form' and '-cline' are different.

Fold attitude

Because of its kinematic significance, the **strike and dip of the axial surface** are important features to measure in an outcrop of a fold. Folds can be classified by axial surface orientation:

Dip	Classification by axial-surface dip
0-10°	Recumbent
10-30°	Gently inclined
30-60°	Moderately inclined
60-80°	Steeply inclined
80-90°	Upright

Because of the way the axial surface is defined, *the hinge of a fold must be a line that lies in the axial surface.* This means that the plunge of the hinge cannot be steeper than the dip of the axial surface. It also means that the hinge line has a rake or pitch in the axial surface. There is a special term for folds in which the rake of the hinge in the axial surface is close to 90°: these are **reclined folds.**

HARMONIC AND DISHARMONIC FOLDS

All the aforementioned qualities are indicative of folds in which each layer is folded in conjunction with the neighboring layers. These are referred to as harmonic folds. Occasionally, we encounter stratified rocks where the folds are misaligned. These are disharmonic structures. Defining axial, inflection, trough, or crest surfaces in disharmonic folds is typically unfeasible, resulting in diminished insights on regional deformation compared to harmonic folds.

Figure 4. This is a rounded, upright, subhorizontal, open, upward facing, synformal syncline. Bramber, Nova Scotia.

Figure 5. Moderately inclined, tight, antiformal anticline. David Thompson Highway, Alberta.

Figure 6. Steeply inclined folds with axial planar cleavage. Carmanville, Newfoundland.

Figure 7. Recumbent fold. Picadilly, Newfoundland.

Figure 8. Variably inclined synform. Port au Port Peninsula, Newfoundland.

Figure 9. Buckle folds in sandstone. Rainy Cove, Nova Scotia.

Figure 10. Kink band in slate. Tancook Island, Nova Scotia.

Figure 11. Refolded isoclinal folds, Pond Point, Bay of Islands, Newfoundland.

Some common types of fold styles

Some types of fold style occur in many circumstances, and can give clues as on the dynamics of folding. Where layers of strong rock are interlayered with very weak rock, the strong layers may have nearly constant thickness around fold hinges, so that the inner and outer arcs are parallel. Such folds are called **parallel** or **concentric** folds. Distinctive styles of parallel folds occur when the strong layers are close together and can slide over each other easily. Under these conditions, very angular folds may develop. When these have inter-limb angles around 60° they are called **chevron** folds. Inter-limb angles around 120° define **kink** folds. Conversely, when all the layers are weak, for example in rocks near their melting point, folds may develop in which each folded surface is geometrically *similar* to the next; these are **similar** folds.

Figure 12. Styles of fold. (a) Parallel folds produced by buckling. (b) Similar folds. (c) Chevron folds. (d) Kink folds.

FOLDS IN MAP VIEW

Folded rocks are more complicated in map view than simple planar strata, because map patterns result from a combination of hills and valleys in the landscape, and antiforms and synforms in the strata.

The shape of folded strata can often be figured out with structure contours, but unlike those for simple planar strata, structure contours for folded strata are not straight. They show angles or curves. Some examples are shown in the following diagrams.

If the fold hinges are perfectly horizontal, the structure contours will remain straight and parallel, but the numbers will show changes of dip:

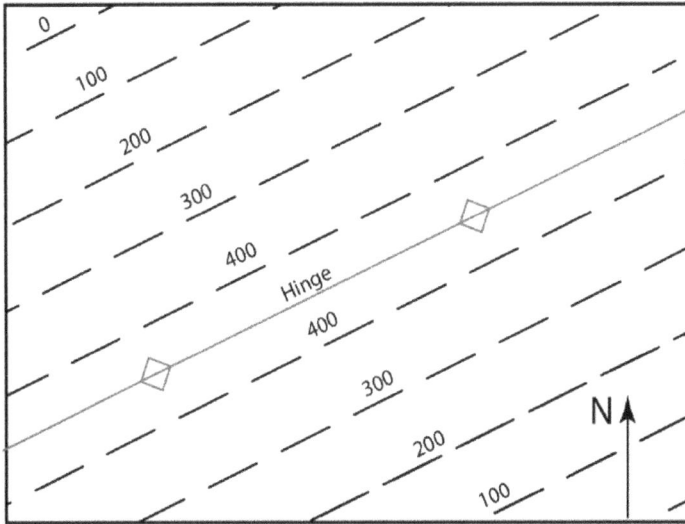

Figure 13. Structure contour pattern for a horizontal antiform.

However, if the fold plunges, the contours themselves will V-shaped, in the case of an angular fold:

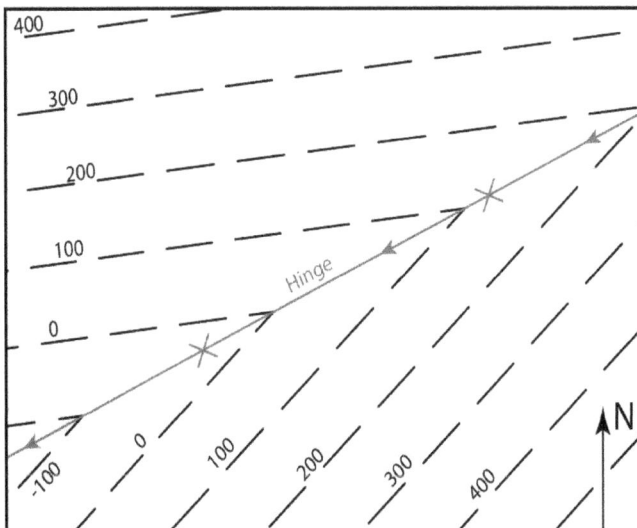

Figure 14. Structure contour pattern for a plunging angular synform.

If the fold is rounded, the structure contours will be curved, showing U-shapes:

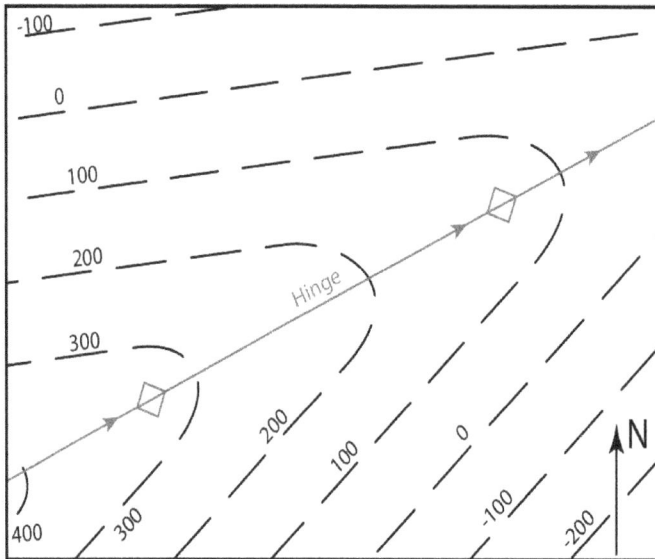

Figure 15. Structure contour pattern for a plunging rounded antiform.

If the fold is overturned, the contours will again be V or U-shaped but contours for both limbs will occur on the same side of the fold hinge:

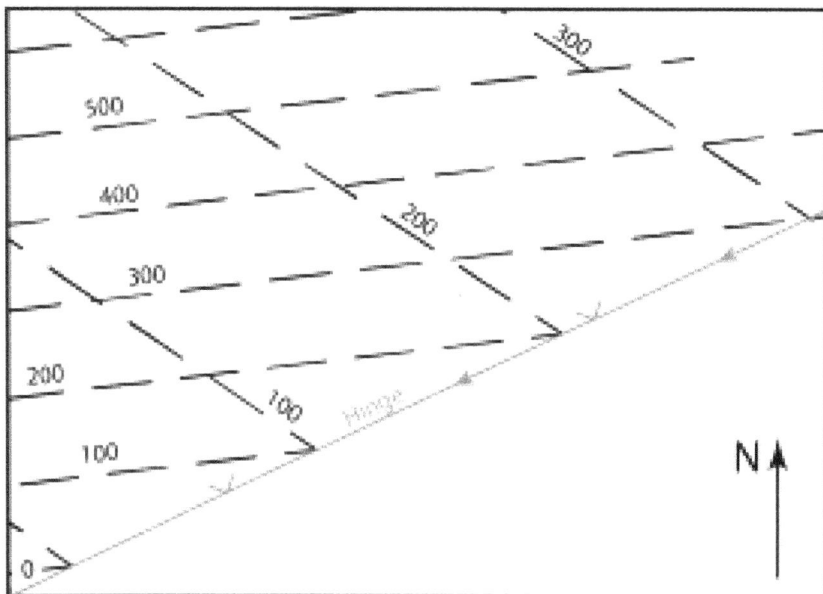

Figure 16. Structure contour pattern for an overturned plunging angular synform.

FOLD OVERPRINTING

Regions subjected to numerous folding occurrences may exhibit refolded structures in a perplexing array of orientations, resulting in diverse fold interference patterns..

Classification of fold interference patterns

To make sense of these we use a classification (devised, like many other fundamental classifications of folds, by J. Ramsay, 1967). This classification relates to two generations of folds only – if we have 3 or more generations things get more complicated.

In each case we characterize a first generation of folds by

- an axial surface
- a fold hinge that lies in that axial surface

We examine the effect of the second generation of folds on those two structures.

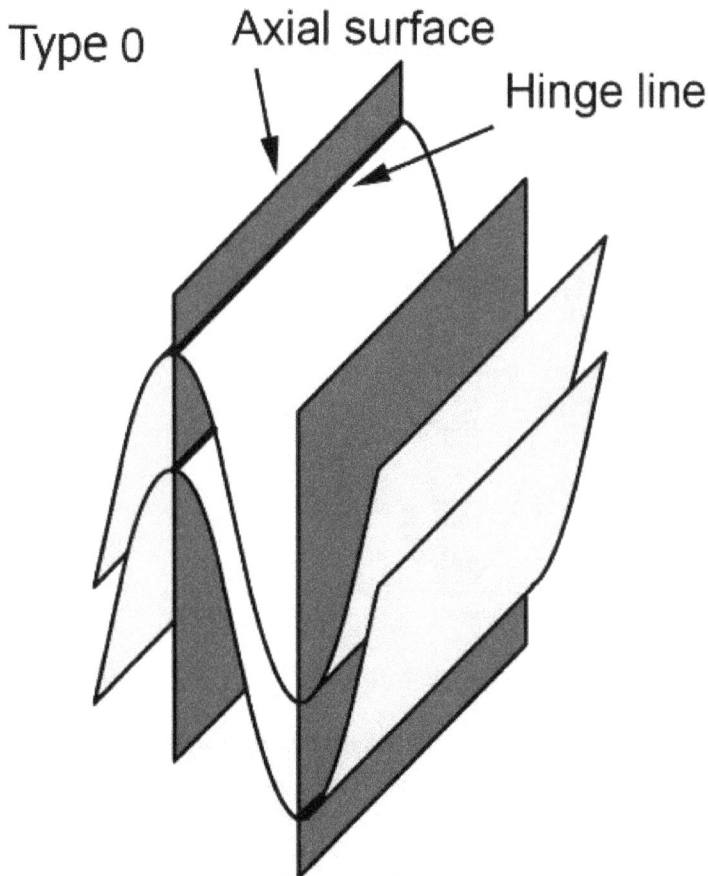

Figure 17. Type 0 fold interference pattern.

If later folding does not bend either the axial surfaces or the hinges of early folds, then the early folds are effectively just tightened during the second phase of deformation. The fold pattern is described as 'type 0', and generally it will not be apparent from geometry that there have been two phases of deformation.

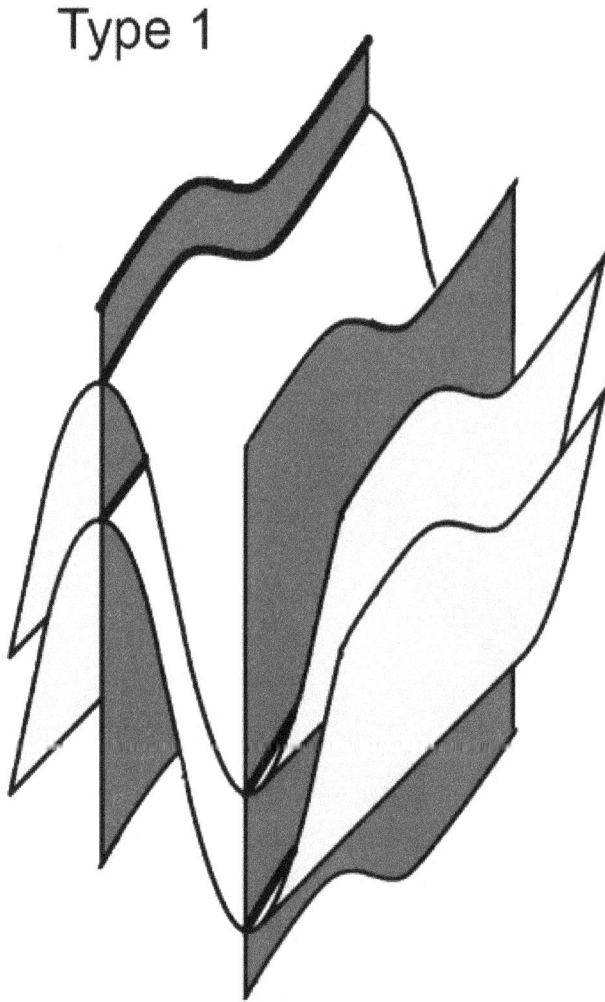

Figure 18. Type 1 fold interference pattern.

If later shortening is roughly parallel to the hinges of the early folds, and the extension direction also lies in the early fold axial surfaces, then later folding bends the hinges of the early folds but does not significantly bend the axial surfaces. This typically produces culminations and depressions on the early fold hinges, and leads to oval outcrop patterns of layer traces on outcrop surfaces. This is called a type 1 fold interference pattern, characterized by domes and basins.

Type 2

Figure 19. Type 2 fold interference pattern.

If on the other hand, the extension direction in the second deformation is at a high angle to the early axial surfaces, then those axial surfaces may be folded too. Under these circumstances outcrop patterns are characterized by complex shapes that resemble mushrooms and bananas. This is called a type 2 interference pattern.

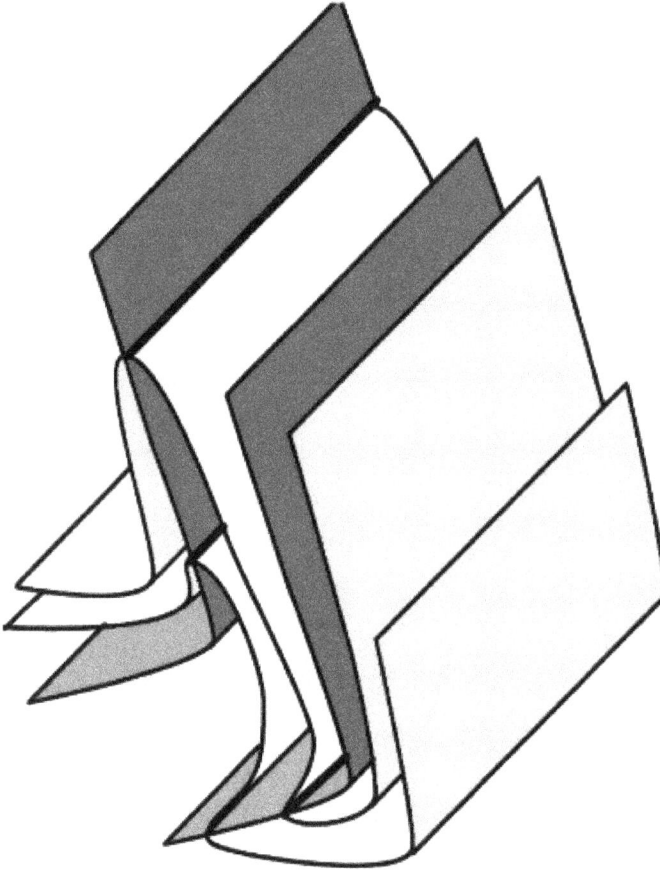

Type 3

Figure 20. Type 3 fold interference pattern.

There is a third case, which occurs when the shortening direction is at a high angle to the early fold hinges and the extension direction is at a high angle to the early axial surfaces. Under these conditions the early hinges may remain more or less straight but the early axial surfaces are folded. The trace of layering typically follows a multiple zigzag shape described as a type 3 interference pattern.

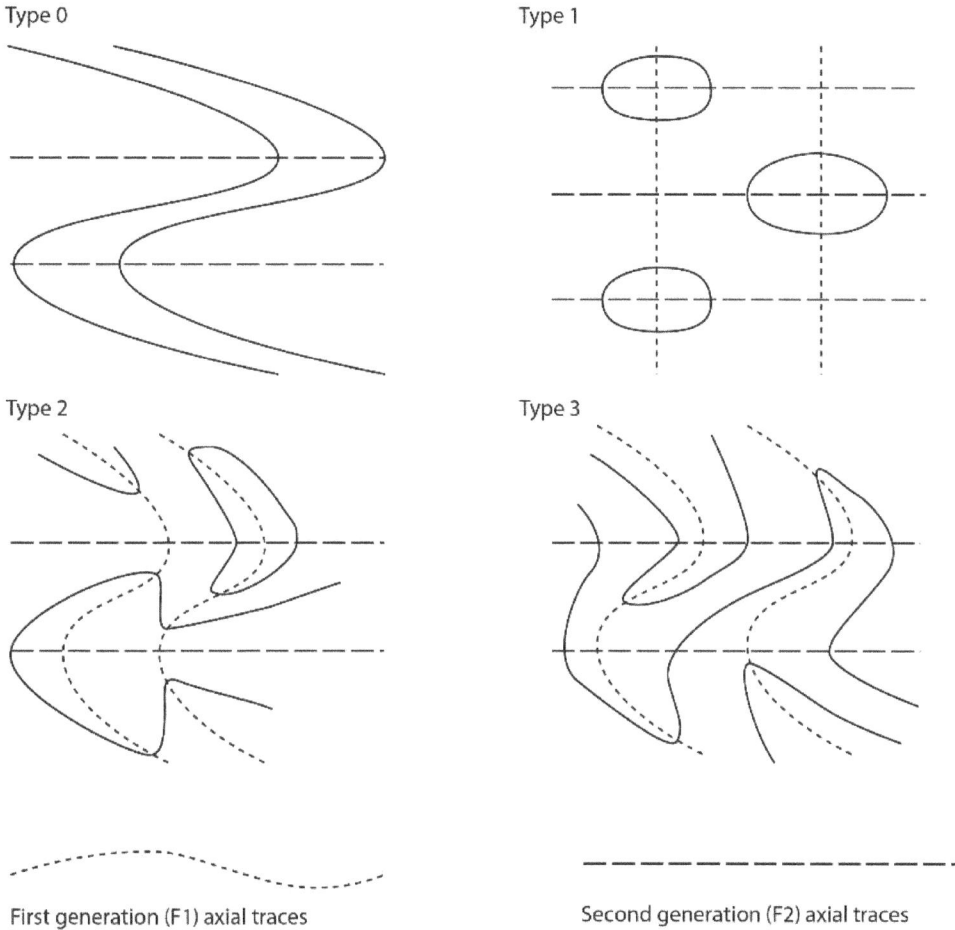

Type 0

Type 1

Type 2

Type 3

First generation (F1) axial traces

Second generation (F2) axial traces

Figure 21. Interpretation of fold interference patterns.

HOW TO ANALYSE AN OVERPRINTED FOLD PATTERN

When trying to understand an overprinted fold pattern, the most important first step is to draw axial traces. In general, the latest set of axial traces is likely to be approximately straight. Earlier axial traces may or may not be folded. (In general, type 1 interference patterns, where early axial surfaces are not folded, are the most difficult to analyze; it may not be possible to determine the order of folding.)

Facing directions may also be useful in the analysis of overprinted folds. For example, downward-facing folds (synformal anticlines and antiformal synclines) always indicate that there are at least two generations of folds.

REFERENCES

Davis, G.H., Reynolds, S.J. and Kluth, C.F. (2011) Structural Geology of Rocks and Regions, 3rd Edition, Wiley, New York, 864 p.

Ramsay, J.G. (1967) Folding and Fracturing of Rocks: San Francisco, McGraw Hill, 568 p.

Van der Pluijm, B.A. and Marshak, S. (2004) Earth Structure: An Introduction to Structural Geology and Tectonics, Norton, New York, 656 p.

LAB 5. MORE ABOUT FOLDS

Multiple planes on the stereographic projection

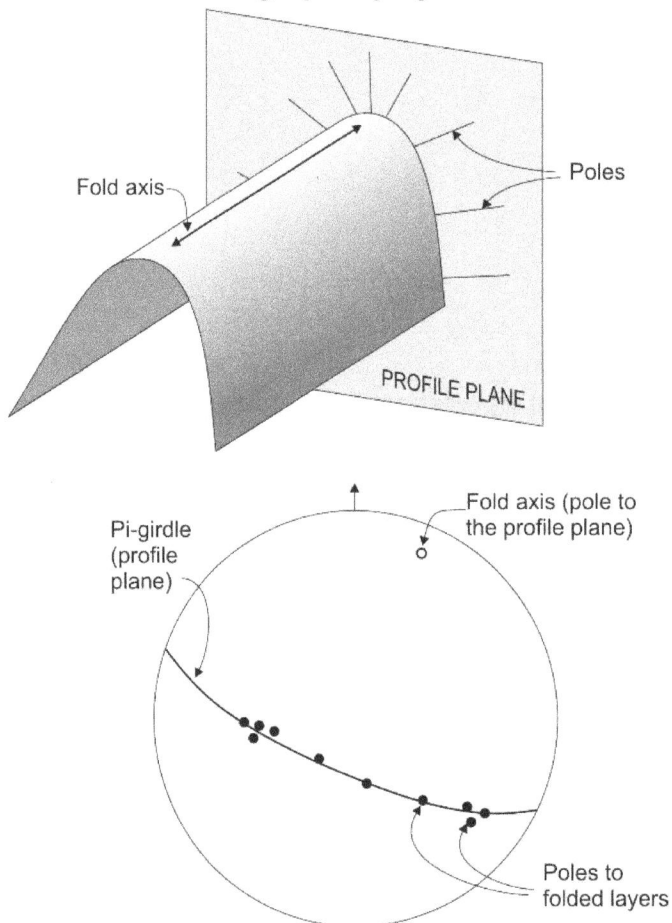

Figure 1. Principle of the pi-diagram, to find the fold axis of a cylindrical fold from measurements of folded surfaces.

The techniques you used in the last lab work well for folds that are perfectly cylindrical, and where the orientations of the folded surfaces are known with great precision. However, in many cases, folded surfaces are not perfectly cylindrical, and small variations in dip, and measurement errors, mean that the geometry is not perfect. Under these circumstances, to determine the orientation of a mean fold axis, it is desirable to measure a large number of orientations, and to use a statistical approach.

Unfortunately, plotting large numbers of great circles rapidly produces a very cluttered projection. For this reason, it is much better to plot **poles** to the folded surfaces rather than great circles. In the ideal case the poles should all lie in the profile plane, and therefore they will plot as points on a single great circle. In real life, things are not so simple; the poles appear scattered in a band, or girdle, on either side of the profile plane, which has to be estimated as a 'best-fit' great circle through the densest band of points.

There is an additional problem, however. If you look at the Wulff net, you will see that the 2° and 10° squares project much larger near the primitive than they do at the centre of the net. This means that plots made with the Wulff net should never be used for statistical inferences based on density of points, because it artificially increases the density in the centre and decreases it on the primitive.

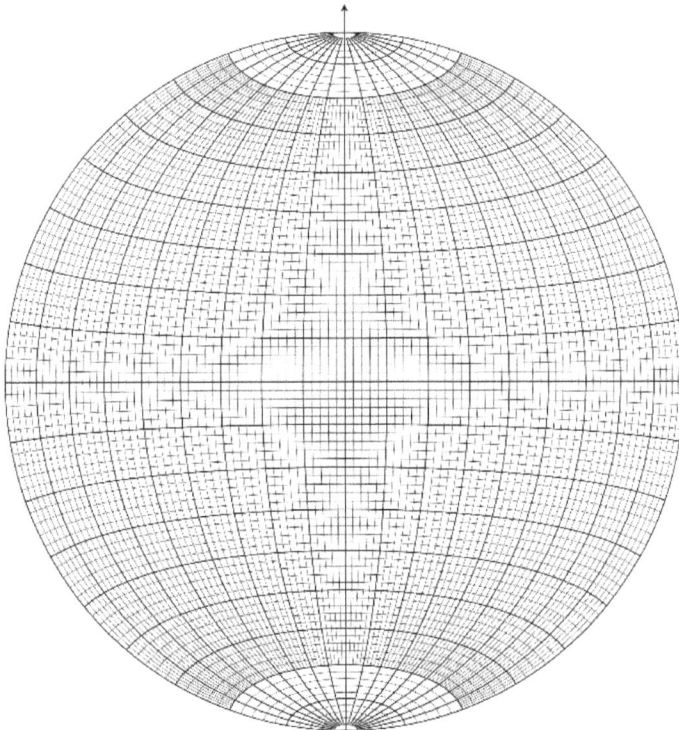

Schmidt net, for plotting equal area projections

For this reason, a modified stereographic projection, called the **equal area projection** is used, based on the **Schmidt net**. All the operations you have met so far are the same on the two nets; the only difference is that the equal area projection does not preserve circular arcs – the great and small circles are complex curves and cannot be drawn with a compass. For this course make sure your net is printed so that its diameter is exactly 15 cm.

Pi diagram

The resulting plot of poles to planes on a equal area projection is called the pi-diagram, which is the most common method used to find the mean fold axis in an area of folded rocks where, as often happens, individual fold hinges are not exposed. The principle of the pi-diagram is illustrated in Figure 1. In a cylindrical fold, the poles to the folded layers are lines perpendicular to the fold axis. They therefore lie in the **profile plane**, the plane that is perpendicular to the fold axis. If the folds are not perfectly cylindrical, or if there are small errors in the measurements, the lines will not lie precisely in the profile plane, but will be close to it.

Contours on the axial surface

Maps of areas with folded rocks can be challenging because of the number of surfaces involved, and because of rapid changes of strike and dip. Often, even though layers may show complex folding, axial surfaces may be approximately planar. Under these circumstances it makes sense to separate the limbs of the fold by drawing axial traces, and even to draw contours on the fold axial surfaces. When doing this, it's important to remember that a single fold will have multiple hinges (one for each folded surface), but that these hinges lie in a single fold axial surface.

Assignment

1. * The area you contoured last week (Great Cavern petroleum prospect) shows strike and dip symbols representing measurements of bedding orientation.

    Plot poles to bedding for all the strike and dip measurements on an equal area projection, to make a pi-diagram. Find and draw the best-fit great circle through the poles; this represents the profile plane. Mark the best estimate of the fold axis (pole to the profile plane) and determine its trend and plunge. Is it similar to the value you obtained by contouring last week, and to the values obtained by your other team members?

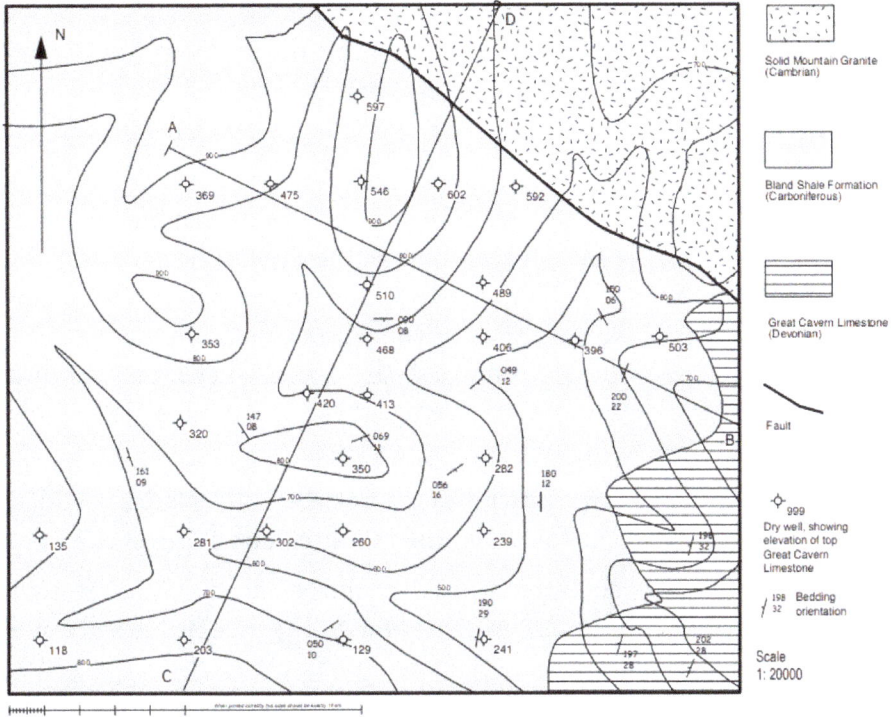

Lab 4 Map 2 Great Cavern Map

2. * Look at the map of Somerset County in the Appalachians of Pennsylvania. The area has a long history of coal mining, and a number of coal seams are identified on the stratigraphic column that doubles as a legend.

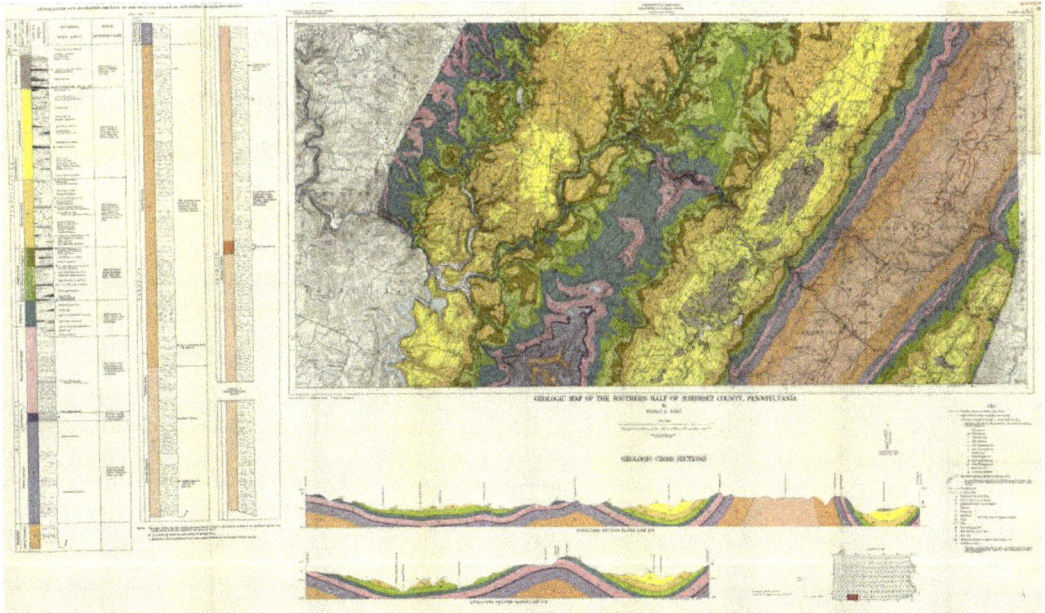

a) The map has structure contours, shown in red, drawn on one surface. Which one surface? (Note: each formation on the map has a top and a bottom surface, so just a formation name is not a complete answer; your answer should be in the form 'the boundary between Formation x and Formation y'.)

b) Use the spacing of the structure contours to determine the strike and dip, at its steepest point, on each limb of the most conspicuous fold. Note that the contours are in feet, and the scale of the map is 1:62500 (about 1 inch to 1 mile). Make sure that in your dip calculation you use the same units vertically and horizontally. Use the points where the contours cross the fold hinge to determine the average trend and plunge of the fold hinge.

c) Plot both limbs of the fold from part 'b' on a stereographic projection. Using the intersection and the angle between the planes, estimate the plunge and trend of the fold axis, and the interlimb angle of the fold.

d) Based on these observations, describe the orientation (plunge, tightness, overall orientation) of the fold in words (e.g. 'tight, steeply plunging synformal anticline')

e) If the fold is cylindrical, the fold axis orientation determined stereographically should coincide with the hinge orientation determined by contours. How close are they (in degrees)?

To find out more, take a look at: Flint 1965 Geology and mineral resources of southern Somerset County, Pennsylvania

3. 3. Map 1 contains an angular fold in a more general orientation than the perfectly horizontal folds you dealt with last week. In addition there are some unfolded rocks and an intrusion. An unconformity separates the more highly deformed folded rocks from the gently dipping younger rocks. To help you solve the map, note that in the west of the map, the bedding traces are somewhat parallel to the topographic contours. These regions correspond to a gentle fold limb. Elsewhere, the geological boundaries cut across the contours at a steeper angle; this is a steep fold limb. Fold hinges can also be identified on the map from sharp swings in the trace of bedding that are not obviously related to valleys and ridges. Before you begin, try to use these hinges to sketch where there might be fold axial traces; these should separate regions of steeper and more gently dipping beds. Do this very lightly – it is likely you will change your mind as you proceed.

    a) Identify the various surfaces on the map. First, mark the unconformity surface in green or yellow. Then mark the boundary between marble and amphibolite in blue or violet. Mark the boundary between amphibolite and schist in red or orange. (Note: the colour scheme is provided for convenience.

If you choose different colours, you must use them consistently throughout this exercise.)

b) Draw structure contours on the green surface.

c) Draw structure contours on the red surface. You will find that there are two parts to this surface: a steep limb and a gentle limb. These have separate sets of contours. The two sets of red contours intersect in a fold hinge. Mark this on the map and trace over it with red.

d) Repeat for the blue surface and mark the blue hinge.

e) The red hinge and the blue hinge are both lines that lie in the axial surface of the fold. Join points of equivalent elevation on the red hinge and blue hinge with lines: these are structure contours on the axial surface. Number them in purple or violet.

f) Use the axial trace contours to predict the outcrop trace of the axial surface, and draw this trace on the map. The axial trace should separate the steep limb from the gentle limb of the fold.

g) Make a cross section along the line XY.

h) Plot both fold limbs, the axial surface, and the unconformity as great circles on an equal area projection. Mark the point corresponding to the orientation of the fold axis.

i) Describe the orientation of the fold in words as completely as you can

j) List the events in the geological history of the area for which you have evidence, starting with the oldest. In the case of the three folded units, where there is no direct evidence of age, you should assume that the fold is upward facing (i.e. synforms are also synclines; antiforms are also anticlines).

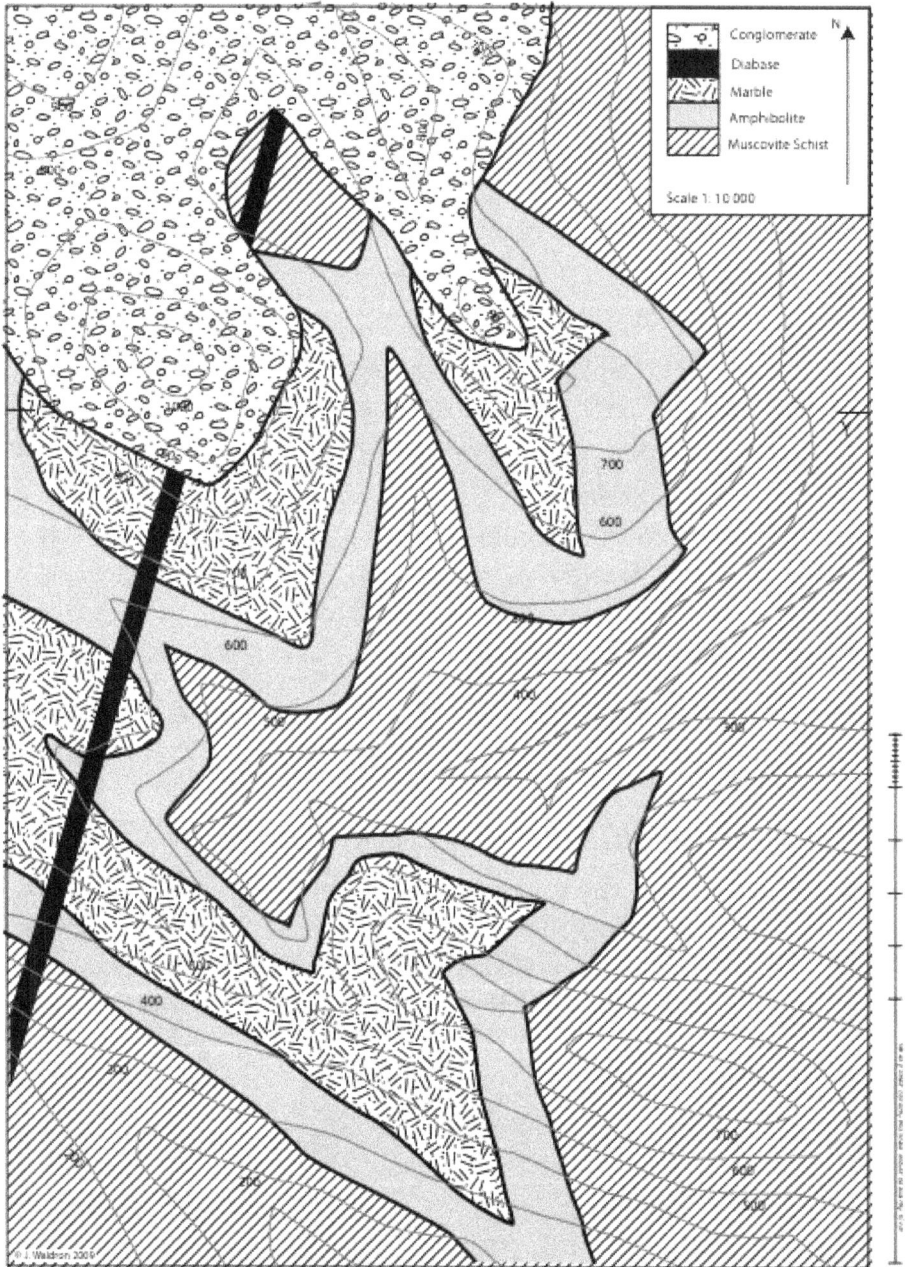

Lab5 Map of Folded Area

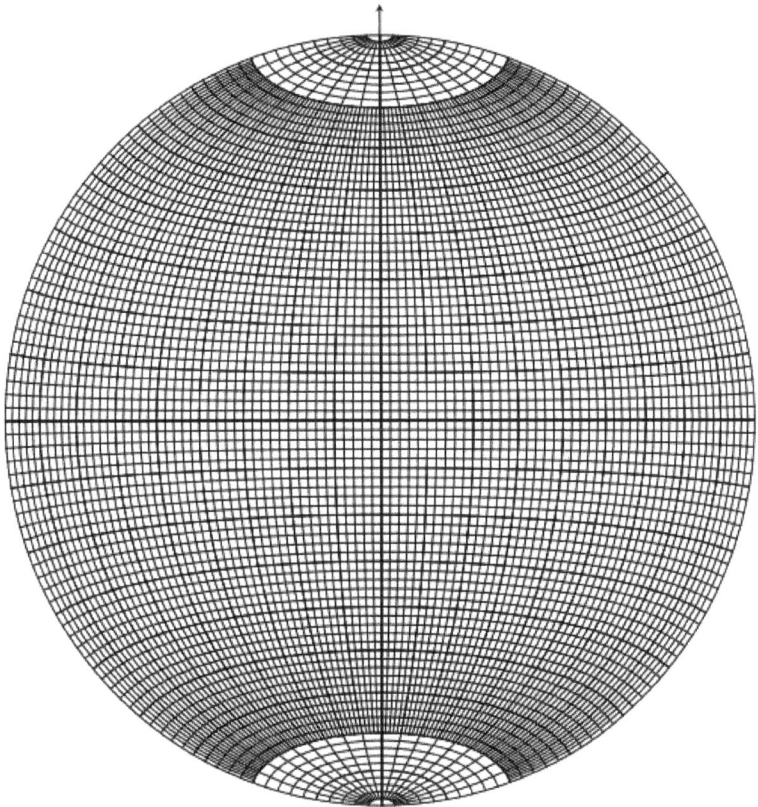

Schmidt Net

Schmidt Net 15 cm

6
Chapter

INTRODUCTION TO BOUDINAGE

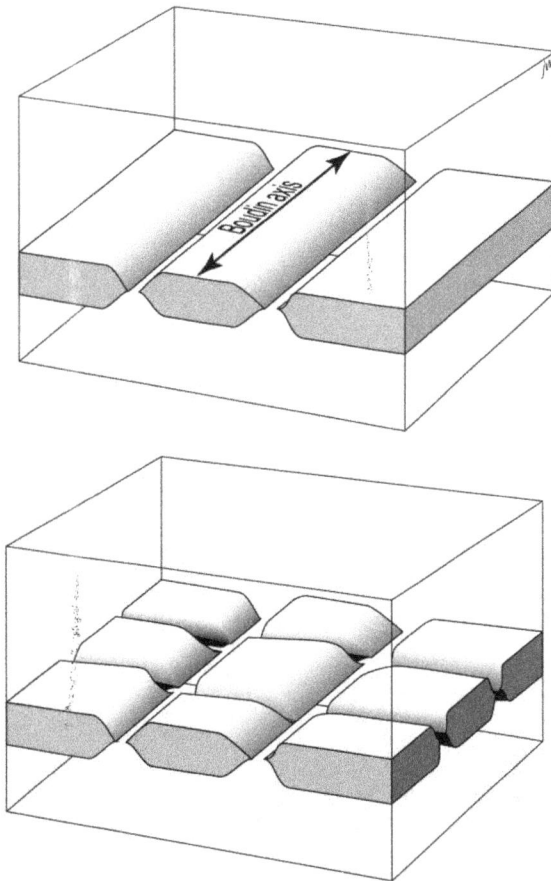

Figure 1. Boudinage. Top: boudins; bottom: chocolate tablet structure.

Buckle folds are formed when strong (or '**competent**') layers of rock are shortened. What happens when strong layers are extended? Typically the layers start to thin at points of weakness (a process known in engineering as **necking)** producing a structure called **pinch-and-swell.** As pinch and swell develops, the thin regions can separate, leaving a structure that looks like a string of sausages in cross-section. The remnants of the original layer are called **boudins** (a French word for a type of sausage), and the process is known as **boudinage.**

Although boudins are in many ways the extensional counterpart of folds, the terminology of boudinage is much less well developed than that of folds. In part, this is because layers undergoing boudinage do not affect adjacent layers in the same way, so that boudins are less likely to be harmonic than folds. Thus, although boudins do have **axes**, it is rarely possible to define an equivalent of an axial surface for boudins.

Sometimes layers undergo extension in all directions simultaneously, producing a more three-dimensional boudinage structure described as **chocolate tablet structure.** It is also possible to find examples of layers that have undergone both folding and boudinage during progressive deformation.

Figure 2. Boudins formed from quartz vein. Carmanville, Newfoundland.

7
Chapter

INTRODUCTION TO KINEMATIC ANALYSIS AND STRAIN

All activities undertaken in this course have focused on delineating the current state of structural geology: the locations of structures and their orientations inside the Earth's crust at now. To comprehend the origin of buildings, it is essential to understand the transformations that occurred throughout their construction, namely the movements involved. The examination of movement throughout geologic time pertains to kinematics.

THE BASIC MOVEMENTS

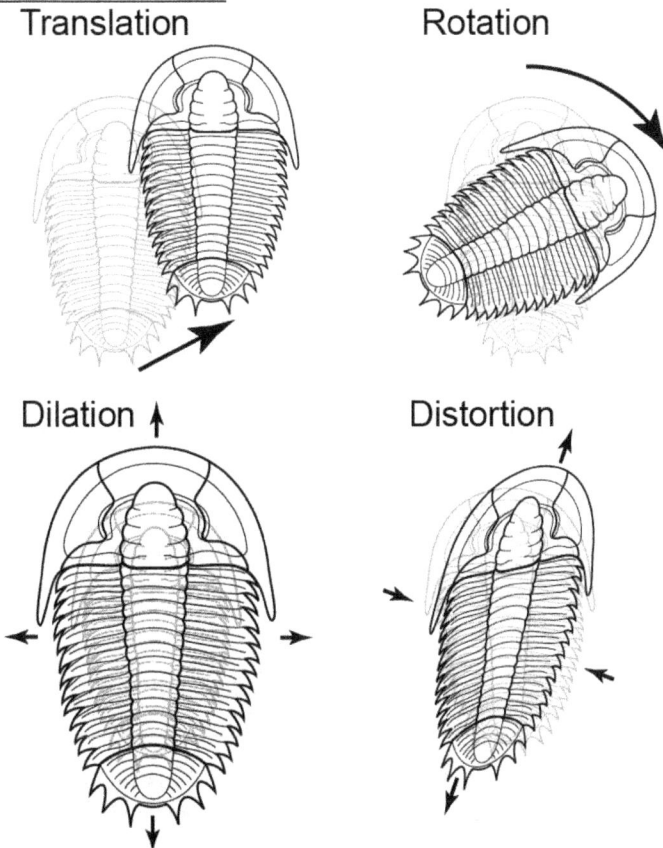

Figure 1. Four parts of deformation.

Deformation can involve 4 types of movement:

- ⊙ Translation, or change in position
- ⊙ Rotation or change in orientation
- ⊙ Dilation or change in area or volume
- ⊙ Distortion or change in shape.

The first two are **rigid body deformations.** The last two together constitute non-rigid deformation or **strain.**

Translation

Translation is measured as a displacement. Any displacement has a distance and a direction. When we look at faults we will measure the displacement of one wall of a fault relative to another. Displacement is a **vector** quantity.

Rotation

Rotation, or change in orientation, is typically measured in degrees, about a particular axis of rotation. (A rotation is therefore also a **vector** quantity, because it has a magnitude and a direction.)

A common kinematic problem involving rotation is to remove the effects of folding from some structure. For example, a sedimentologist may measure the orientation of a paleocurrent structure in folded strata, and wants to know the original direction of current flow.

This type of problem is most easily solved on the stereographic projection. As beds are 'unfolded', the paleocurrent directions they contain rotate along small circles on the stereonet. The small circles are centred on the fold axis.

Dilation

Volume change is very difficult to measure in real rocks. However, under some circumstances it can be quantified. Most sedimentary rocks undergo some **compaction** as they are buried, because pore water is expelled. Compaction results in a negative dilation. Another common phenomenon involving negative dilation is **pressure solution** in which some minerals in a rock are dissolved in response to stress.

Dilation is a **scalar** quantity – it just has a magnitude.

Distortion

Distortion is by far the most complicated type of deformation to measure. When rocks are distorted they typically get longer in some directions and shorter in others. Also, angles change in distortion. Because of this, strain cannot be represented by a scalar or a vector. It is a more complicated quantity that is called a **tensor.**

Strain

In common vernacular, the term "strain" evokes the notion of exerting power; "straining" at something suggests a dynamic activity. The term "strain" possesses a distinct definition in the realms of science and engineering, referring specifically to a kinematic alteration in shape or volume, irrespective of the cause. There are instances of strain that do not entail any force. A reflection in a distorting mirror and the warped picture of a basketball on a widescreen television are both representations that have experienced distortion, yet they consist solely of light rays; no external force is responsible for the distortion.

(The corresponding dynamic term is "stress", which we will study in a later section. Take care never to confuse stress with strain!)

All structures observed in the Earth's crust are representations of strain.Strain is a key notion in structural geology, leading to its precedence in mathematical treatment in advanced courses.This course has a pragmatic approach, prioritizing the description of structures first.It should now be evident that alterations in the morphology of rocks can be identified in all the structures you have examined to now.

Heterogeneous strain and homogeneous strain

Strain can vary from place to place in rocks. We distinguish a special case:

Homogenous strain: strain that is the same everywhere within a body of rock. In homogeneous strain, straight lines remain straight, parallel lines remain parallel, circles are deformed into *ellipses.*

Strain that is not homogeneous is **heterogenous.** In heterogenous strain, straight lines can be bent, and lines that were initially parallel are rotated by different amounts, becoming non-parallel.

Heterogeneous strain is difficult to deal with mathematically. However, if we look at a very small region of a heterogeneously strained rock, it can often be treated in the same way as a homogeneous strain. The **strain at a point** in a heterogeneously strained rock follows the same mathematical rules as homogenous strain.

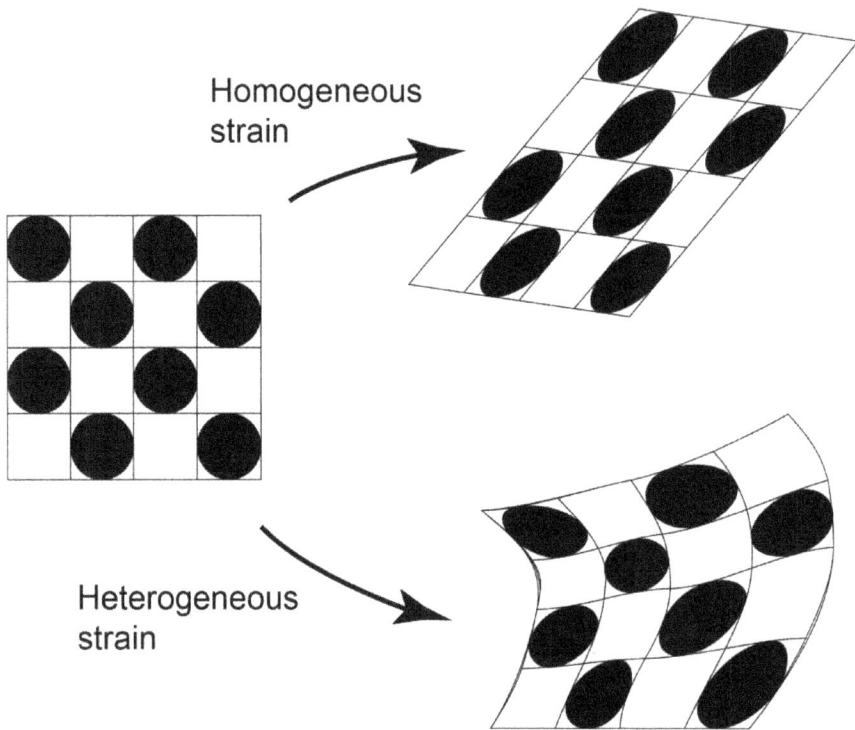

Figure 2. Homogeneous and heterogeneous strain.

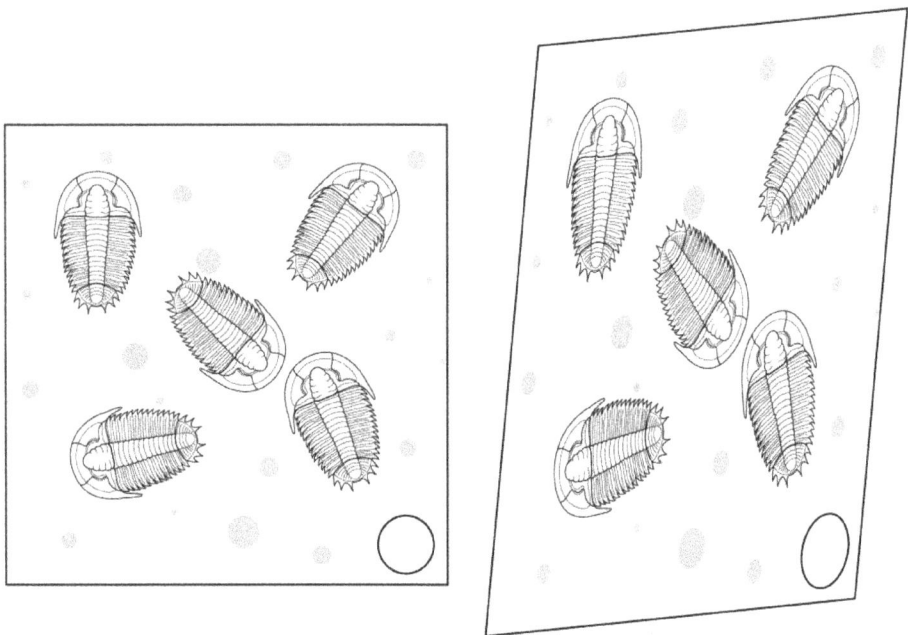

Figure 3. Strain affecting variably oriented trilobites. Note that each fossil is affected differently, depending on its initial orientation, but the strain is nonetheless homogeneous, as shown by the ellipses. Trilobite image after Gon, S.; www.trilobites.info; Downloaded 2012 Dec 17.

Strain along lines

In a homogeneous strain, lines orientated differently undergo varying degrees of strain.In the diagram, each trilobite experiences varying effects from deformation, despite the total strain being uniform.

CHANGES IN LENGTH (LONGITUDINAL STRAIN)

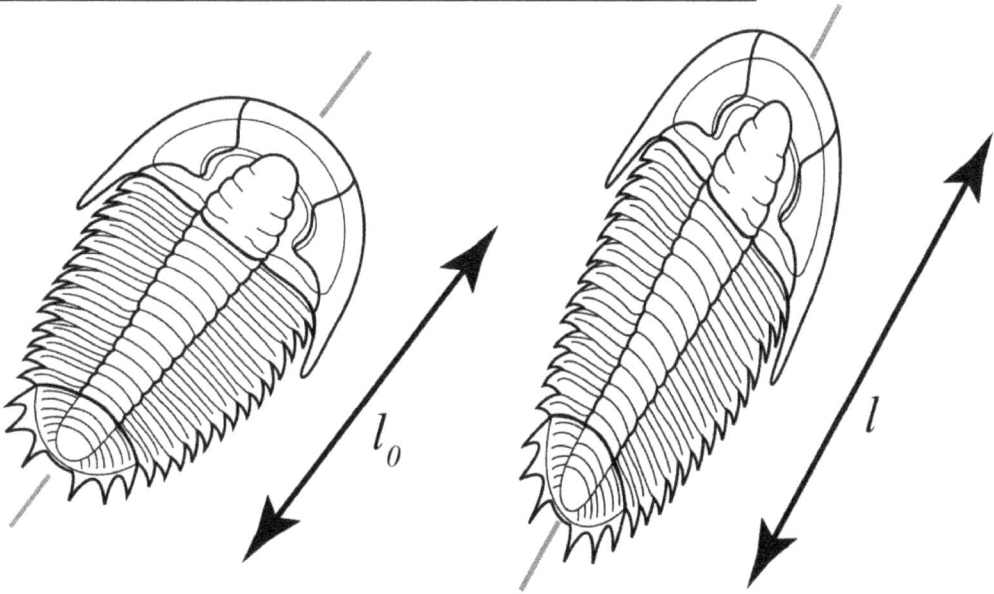

Figure 4. Longitudinal strain as it affects the mid-line of one trilobite in Figure 3.

There are two ways to measure change in length. If original length is l_0 and new length is l,

Extension (sometimes elongation)

$e = (l-l_0)/l_0$

Extension is the fractional change in length.

Stretch $s = l/l_0 = 1+e$

The important thing to remember is that in strained rocks the elongation varies with direction: typically some lines will have got longer and others will have got shorter.

Changes in angles (shear strain)

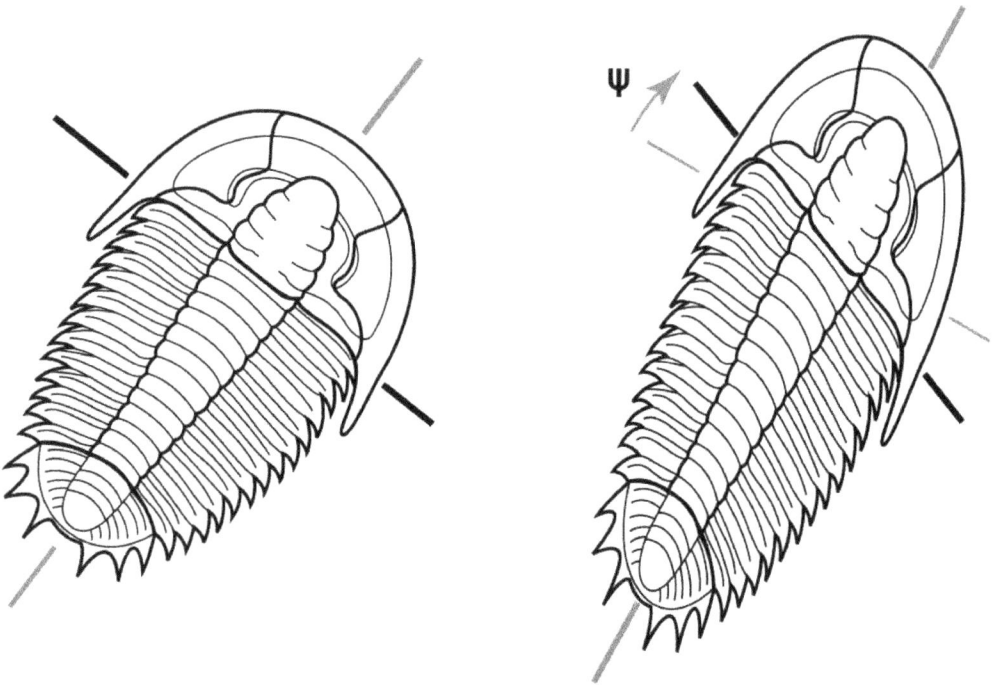

Figure 5. Shear strain as it affects the mid-line of one trilobite in Fig. 3.

To measure change in angles we look at two lines that were originally perpendicular.

If the change in angle is ψ (psi) then

Shear strain (gamma) γ= *tan* ψ

There are other methods used to measure shear strain in advanced studies, so the above definition is sometimes qualified as **engineering shear strain** because it is the measure allegedly most used by engineers.

Shear strain also varies with direction – some lines undergo positive shear strain and some undergo negative shear strain.

In the diagram, the *long axis* of the trilobite has undergone *positive shear strain* because the line originally perpendicular to it has rotated *relatively clockwise*.

Strain ellipse

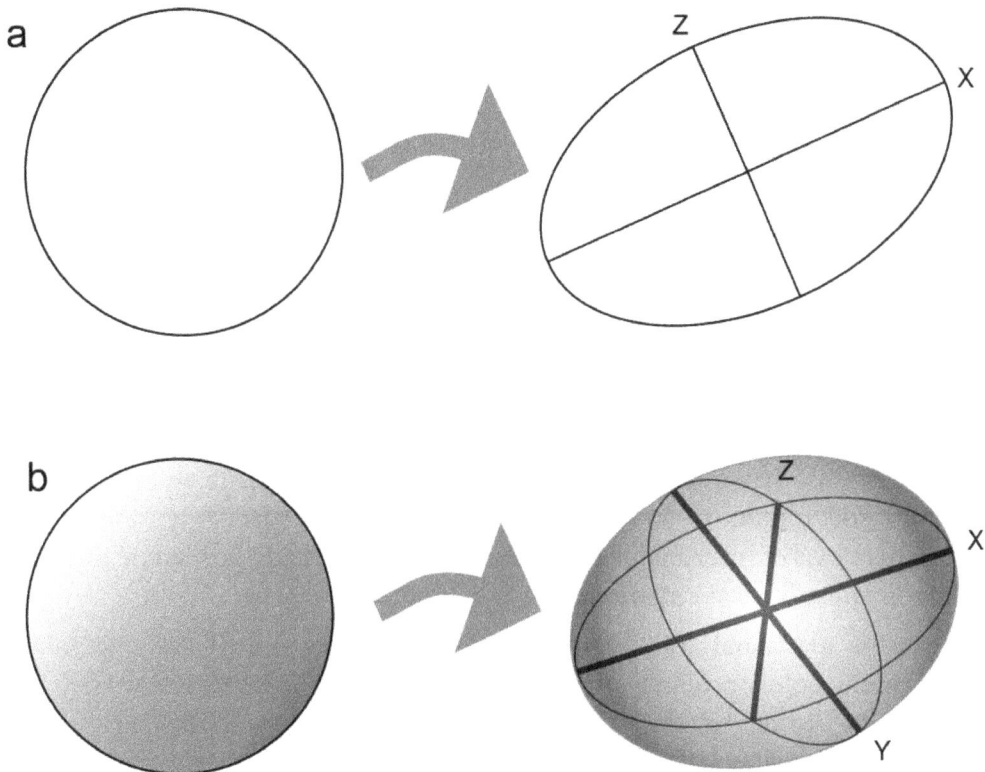

Figure 6. (a) Strain ellipse; (b) strain ellipsoid.

The strain ellipse effectively illustrates a state of homogeneous strain or the strain at a specific point on a two-dimensional plane. The strain ellipse represents the configuration of a deformed circle that initially possessed a unit radius. The radius of the strain ellipse in any direction corresponds to the stretch s in that direction. The strain ellipse effectively illustrates the fluctuation of longitudinal strain with direction.

Strain axes

A strain ellipse has two lines that are special. They represent the maximum and minimum stretches. These lines are **strain axes**. (The strain axes are sometimes called X and Z. (Note: coordinate axes on a map or cross-section are sometimes x and z. Strain axes don't necessarily coincide with any special direction on a map.)

Strain axes have some other special properties. They are always at right angles to each other, and they also represent lines of *zero shear strain.* This means that they were perpendicular before deformation started too. (However, during deformation they may have diverged from this perpendicular relationship and then come back to it!). The strain axes are the only lines that have this property.

The **strain ratio** is a convenient measure of the amount of distortion in 2-D. The strain ratio is the ratio between the long axis and the short axis of the strain ellipse:

Strain ratio $R_s = s_x/s_z$

Strain ellipsoid

The **strain ellipsoid** is a convenient way to represent a state of homogenous strain, or the strain at a point, *in three dimensions.* The strain ellipsoid is the shape of a deformed sphere that originally had unit radius.

The radius of the strain ellipsoid in any direction is equal to the *stretch s* in that direction. In 3-D, the strain ellipsoid is thus a good way to represent the variation of longitudinal strain with direction.

A strain ellipsoid has *three* lines X,Y and Z that are special. They represent the maximum and minimum stretches, called s_X and s_Z respectively, and a third, intermediate axis of intermediate stretch s_Y, that is mutually perpendicular to the other two. These lines are **strain axes**.

They possess additional unique qualities. They are perpendicular to one another and represent poles to planes of null shear strain.This indicates that they were perpendicular prior to the onset of deformation as well.However, during deformation, they may have deviated from this perpendicular relationship before returning to it.The strain axes are the sole lines possessing this feature.

In 3D the strain axes are poles to three **principal planes of strain.**

In 3D the shape of the strain ellipsoid can't be defined by a single strain ratio. Instead we recognize two strain ratios $a = s_X/s_Y$ and $b = s_Y/s_Z$. If a is large and b is small then the strain ellipsoid looks like a football or a cigar, and is described as **prolate.** Rocks with prolate strains often display strong linear fabrics (lineations). In contrast if a is small and b is large then the strain ellipsoid looks like a cushion or a pancake, and is described as **oblate.** Rocks with oblate strains often display strong planar fabrics (foliations).

Prolate ellipsoid
("cigar")

Plane-strain ellipsoid
(neither oblate nor
prolate)

Oblate ellipsoid
("pancake")

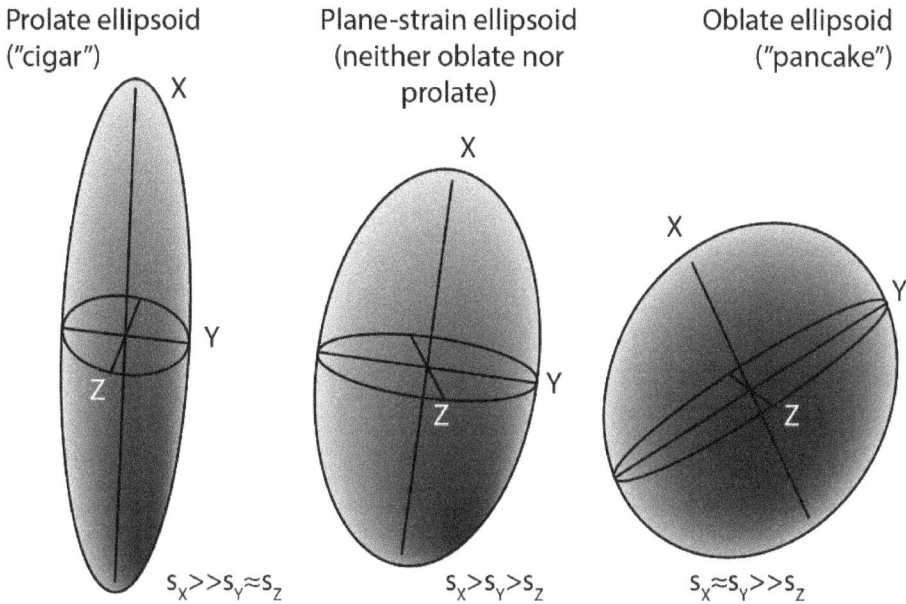

$s_X \gg s_Y \approx s_Z$

$s_X > s_Y > s_Z$

$s_X \approx s_Y \gg s_Z$

Figure 7: A range of strain ellipsoid shapes at constant volume.

The range of possible strain ellipsoid shapes can be illustrated on a graph of a against b known as a Flinn plot.

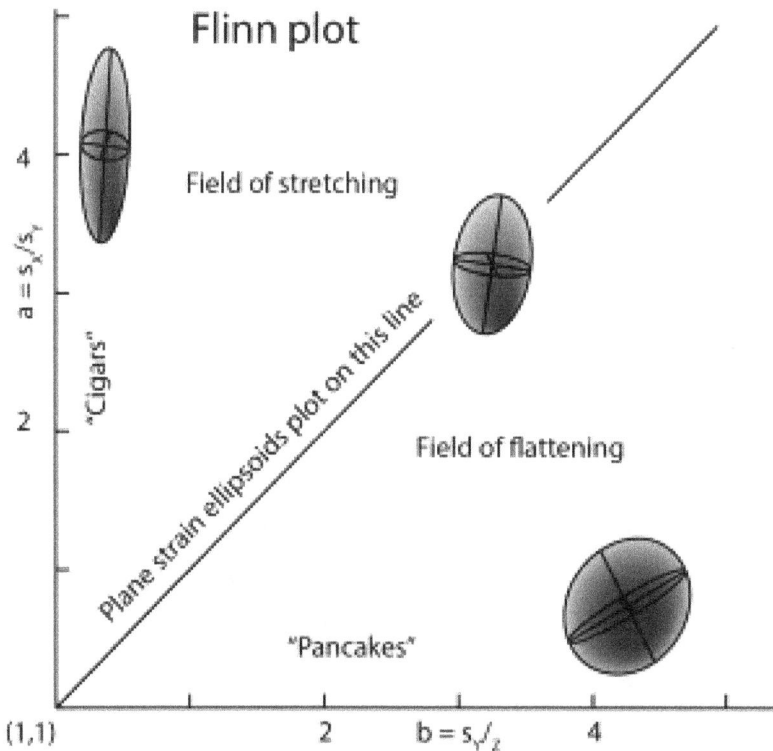

Figure 8: Flinn plot for constant-volume strain ellipsoids

DEFORMATION HISTORIES

Rotational and non-rotational deformation

All the above measures have concerned just strain. However, if we look at the whole deformation picture, we may see situations where rotation has gone on at the same time as strain. Under these circumstances it's helpful to look at the behaviour of the strain axes.

If the strain axes have the same orientation as they did before deformation started, then the deformation is **non-rotational** (sometimes **irrotational**). (Note that even in a non-rotational strain other lines will have been rotated, and will show shear strain, but it is the strain axes that are important for the terminology here.)

If the strain axes have rotated during deformation, then the deformation is described as **rotational.**

Finite deformation and deformation rate

Once we start looking at rotation, it's difficult to avoid discussing **strain history** too. When we look at a deformed rock what we see is the product of a whole history of deformation. That end product is called the **finite deformation** and the strain part of the deformation is the **finite strain**.

In more detailed analyses of strain, we may be interested in all the tiny **increments** of strain that have contributed to the final picture. Each one is called an **incremental strain.**

The **infinitesimal strain** is the end product of this type of thinking. The strain history is thought of as being made up of an infinite number of infinitesimal strain increments. (The general idea should be familiar if you have taken a class in calculus.) This idea of infinitesimal strain becomes important if we look at **strain rates.** Strain rates are typically measured in units of **per second,** or s^{-1} sometimes expressed as **strains per second.** In geologically reasonable situations, the amount of strain that occurs in a second is almost infinitesimal. Typical ductile strain rates in the Earth's crust are thought to be between 10^{-12} and 10^{-15} s^{-1}.

COAXIAL AND NONCOAXIAL DEFORMATION HISTORIES

If the strain increments are all non-rotational, then the strain axes maintain their same direction relative to the material of the rock. The deformation is called **coaxial.** This type of deformation is also called **pure strain.**

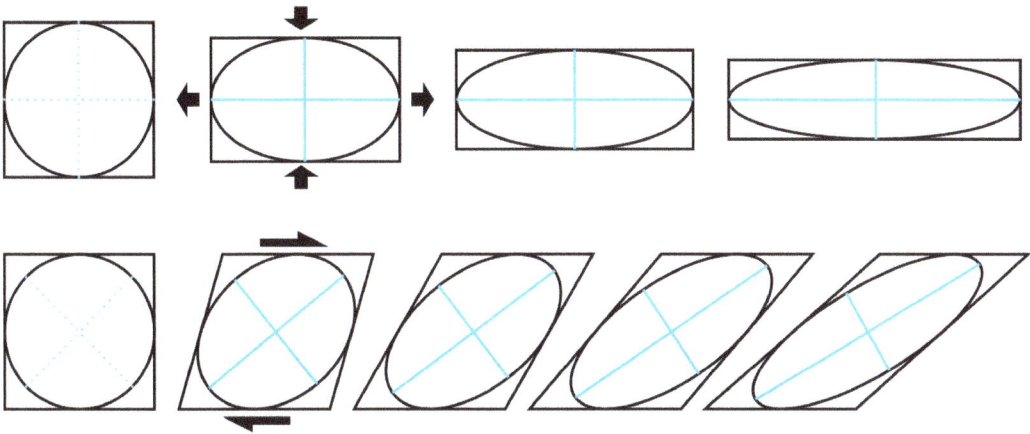

Figure 9. Top: Coaxial or pure strain. Note that the strain axes keep their directions constant relative to the material being deformed. Bottom: Simple shear, one type of noncoaxial strain history. Note that the strain axes rotate over time relative to the material being deformed.

Any other type of strain history is a **noncoaxial.** There are many types of noncoaxial strain history, but one particular type is important in the study of shear zones. It is **simple shear**. In simple shear, all the particles of the rock move in the same direction, but the strain axes are progressively rotated. Sometimes, by looking in detail at the fabric of shear zones, we can distinguish the sense of rotation and figure out which way the rocks were moving.

8
Chapter

INTRODUCTION TO FABRICS

INTRODUCTION: AN OVERVIEW

Any rock exhibiting directional variability in its characteristics is described as having fabric. Similar to most buildings, fabrics can be classified as primary or secondary. Instances of primary fabric include layering in sedimentary rocks and flow banding in igneous rocks. Secondary textiles generally signify that deformation has taken place. Fabrics frequently occur in folded rocks, and when they form concurrently with folding, they typically exhibit a distinct relationship to the folds. Fabrics can be classified as planar or linear. Planar textiles are referred to as foliations. Some textbooks limit the term foliation to specific forms of planar fabric, such as those with coarse mineral grains seen in schist; nevertheless, we shall employ it in a broader context to refer to any planar fabric. Rocks exhibiting a pronounced secondary planar fabric are referred to as S-tectonites, with the 'S' denoting schistosity. Linear fabrics are referred to as lineations. Rocks exhibiting pronounced secondary lineation are referred to as L-tectonites. A lineation often occurs parallel to, or inside the plane of, a planar fabric. Rocks exhibiting a pronounced tectonic lineation aligned with the plane of a tectonic foliation are referred to be LS-tectonites.

FABRIC ELEMENTS

The concept of a fabric element is essential to the description of fabrics.Fabric elements (Fig. 1, 2) are the components arranged to impart a fabric to a rock.Fabric elements encompass layers and lenses of varying composition, as well as tabular mineral grains (such as mica) and acicular mineral grains (such as hornblende or sillimanite).Folds may also constitute fabric elements, provided there is a sufficient quantity of them.This introduces a second significant concept: the notion of a penetrative fabric. A fabric is considered penetrative if it is ubiquitous within the rock.For instance, certain slates can be divided at nearly any location, resulting in a virtually limitless number of cleavage planes.In other instances, the cleavage planes are distinctly separated and are referred to as non-penetrative.

Penetrative fabrics

A fabric is **penetrative** if:

- ⊙ It is present throughout the rock;
- ⊙ It is not possible to count the number of fabric planes or lines;
- ⊙ It is not possible to see spaces between the fabric planes or lines.

Non-penetrative fabrics

A fabric is **non-penetrative** if:

- ⊙ The fabric planes or lines are spaced apart within the rock, or
- ⊙ It is possible to see spaces between the fabric planes or lines where no fabric is present, or
- ⊙ The fabric planes or lines can be counted.

Note that penetrative character is a scale-dependent concept; a fabric that is penetrative at map and outcrop scale may prove to be non-penetrative in thin section.

FABRIC ELEMENTS

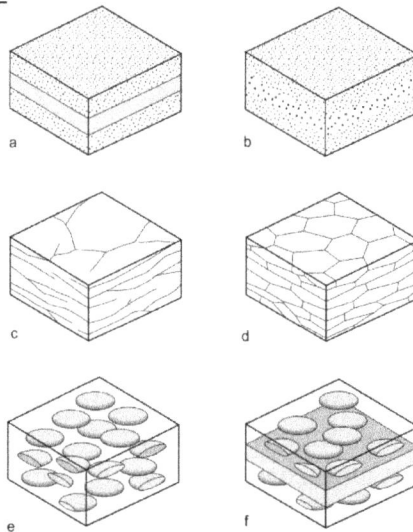

Figure 1. Different types of foliation and their fabric elements. In each case the foliation is parallel to the top surface of the block. (a) Compositional layering; (b) grain size variation; (c) closely-spaced fractures; (d) preferred orientation of grain boundaries; (e) preferred orientation of domains of different composition; these domains could be mineral grains, groups of mineral grains, deformed pebbles, etc. These various types of fabric element can be combined. (f) A combination (a + e) that is common in sedimentary and metamorphic rocks. Based on Hobbs, B.E., Means, W.D., and Williams, P.F. 1976. An Outline of Structural Geology. Wiley.

The following types of fabric elements can give a rock a fabric. Notice that it's common for a given fabric to be defined by a combination of fabric elements, as shown in Figures 1 and 4.

Tabular mineral grains:

Platy or flake-shaped mineral grains like mica are often aligned to produce a foliation.Less obviously, tabular mineral grains may contribute to lineation if their orientations are scattered so that their planes all include a particular line.

Acicular mineral grains:

Needle-shaped, or acicular, mineral grains like amphibole and sillimanite are often aligned to produce a lineation.Less obviously, needle-shaped mineral grains may contribute to foliation if they are distributed parallel to a plane.

Crystallographic preferred orientation (CPO):

Occasionally, it is essential to examine a thin segment under the microscope to observe a fabric. If the crystallographic axes of a specific mineral are aligned with a particular plane or line, the rock exhibits a fabric.This phenomenon typically becomes apparent when the microscope stage is rotated between crossed polarizers, particularly with the insertion of a 1lambda plate; all grains exhibit synchronized color changes as the stage rotates.

Domains:

A domain is a region within a rock that has a distinctive composition or texture. Flattened clasts in a sedimentary rock can produce a foliation if they are all aligned to a plane. Stretched clasts in a sedimentary rock can produce a lineation if they are all aligned parallel to a line.

Layers:

Layers are really just domains that are very extensive and parallel-sided.

Rods:

Rods are really just domains that are very elongated and continuous.

Cracks and discontinuities:

Closely spaced fractures can produce a type of cleavage called **fracture cleavage**.

Fold hinges:

Numerous closely spaced fold hinges may produce a lineation in highly deformed rocks. This structure is called a **crenulation lineation**.

Fold axial surfaces:

Numerous closely spaced fold axial surfaces define a fabric called called **crenulation cleavage**, because the axial surfaces may be planes of weakness along which a rock tends to split.

Intersecting foliations:

In the presence of two intersecting foliations, an intersection lineation is invariably defined by their intersection. While it is feasible to compute the intersecting lineation using strike-and-dip measurements of the two foliations in the field, direct measurement of the lineation is generally more precise.

Common types of foliation, and their origins

Figure 2. Sedimentary foliation: bedding and fissility, Arisaig, NL.

Primary fabrics:

Sedimentary rocks are typically recognized by their primary fabrics. Actually, two types of fabric are common:

1. **Bedding** and **lamination** are defined by layers of different composition or texture, typically reflecting environmental change during deposition;
2. **Fissility** is a fabric characteristic of mudrocks, in which the sheet silicates have been aligned by **compaction** to produce a penetrative primary foliation.

Figure 3. Axial planar slaty cleavage parallel to hammer handle, Carmanville NL.

Slaty cleavage:

Slaty cleavage is a penetrative texture in which micas or other sheet silicates, imperceptible to the human eye, are oriented parallel to a plane, creating a preferential plane of separation.Research on strain has demonstrated that slaty cleavage generally develops orthogonal to the axis of maximal shortening, which is the Z axis of the strain ellipsoid.Cleavage occurs parallel to the X-Y plane. Certain slaty cleavages penetrate to the level of individual mineral grains, while others exhibit domains under microscopic examination.

Cleavage is thought to develop by a combination of processes including:

1. Physical rotation (transposition) of mineral grains as a rock is deformed;
2. Solution of grains subjected to high stress (pressure solution);
3. Growth of new mineral grains during deformation.

Slaty cleavage is the defining characteristic of the rock-type **slate**.

Schistosity:

Schistosity is a coarser grained version of slaty cleavage, and is the term used for a more or less penetrative foliation defined by mineral grains coarser than about 1 mm. Many geologists actually recognize an intermediate feature **phyllitic foliation**, when the fabric-defining minerals are just visible. However, the boundaries between these categories have never been formally defined.

Schistosity involves the same types of process as slaty cleavage, but is typical of higher metamorphic grades and always involves significant new mineral growth.

Gneissic banding:

Gneissic banding forms due to recrystallization during medium- to high-grade metamorphism. It is defined by the favored alignment of platy, tabular, or prismatic minerals, as well as by subparallel lenticular mineral grains and aggregates. Most gneisses exhibit compositional (mafic/felsic) banding.

Flattening fabric:

When a foliation is characterized by domains that signify recognizable deformed entities inside a rock, such as compressed stones in a deformed conglomerate, the fabric is referred to as a flattening fabric.Most foliations are likely generated by flattening, although this word is applied when the origin s evident.

Pressure-solution cleavage:

Sometimes deformed rocks contain domains or layers of different composition that can be shown to have originated as a result of solution of particular minerals. For example, quartz may be preferentially dissolved from certain bands within a rock, producing mica-rich **seams** that separate quartz-rich **lithons**.

Figure 4. Pressure-solution cleavage, Bay of Islands NL.

Crenulation cleavage:

Crenulation cleavage is defined by closely spaced fold axial surfaces. It's almost always a second or later generation of fabric: an initial deformation produces a foliation and later deformation folds that foliation to produce a crenulation cleavage.Sometimes crenulation cleavage is combined with pressure solution to produce **differentiated crenulation cleavage**.

Mylonitic foliation:

Extreme ductile shearing in shear zones tends to reduce the grain size of mineral grains, while stretching the original grains out into domains of extreme dimensions, and also producing strong crystallographic preferred orientation. This type of foliation is called **mylonitic** foliation. It's almost always accompanied by a lineation, producing an LS-fabric.

RELATIONSHIPS OF FOLIATIONS TO FOLDS

Foliations are often developed parallel, or roughly parallel, to the axial surfaces of folds (Fig. 3). This relationship is common because both structures originate from the same strain. Fabric planes and fold axial surfaces are perpendicular to the shortening direction, the short axis of the strain ellipsoid (Z). Foliation of this type is called **axial planar foliation** (or axial planar cleavage, etc. depending on the type of foliation.)

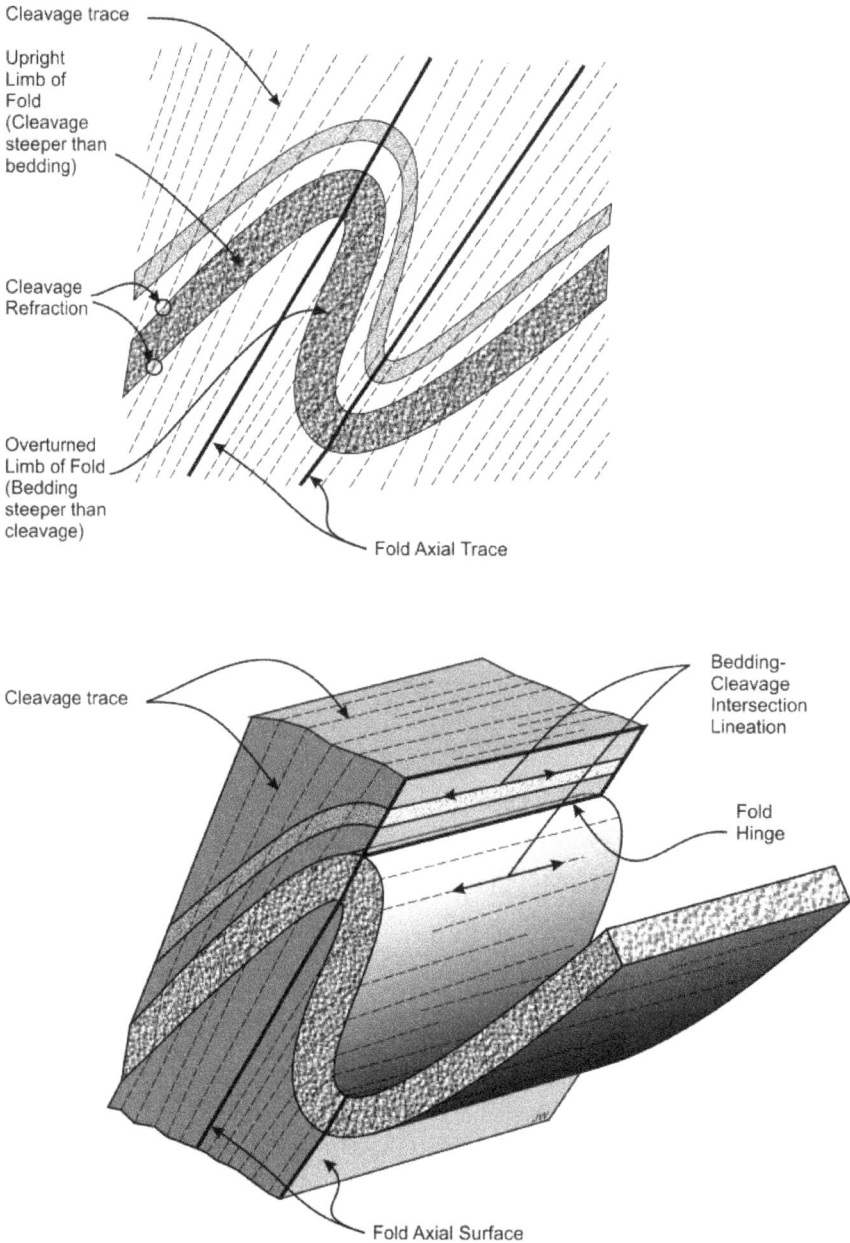

Figure 5. Common relationships between folds and fabrics.

In detail, axial planar foliations often depart from exact parallelism with the axial surface of a fold, because of local strain variation. One very common style is called **cleavage refraction** (Fig. 5) in which cleavage bends so that it is more perpendicular to competent layers, and more parallel to incompetent layers.

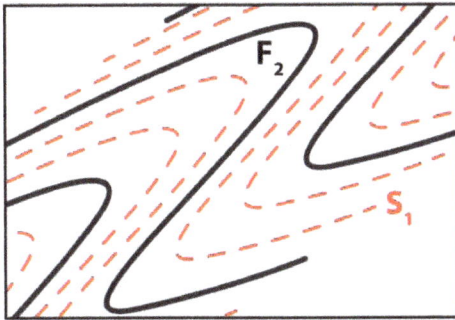

Fabric S_1 predates folds F_2

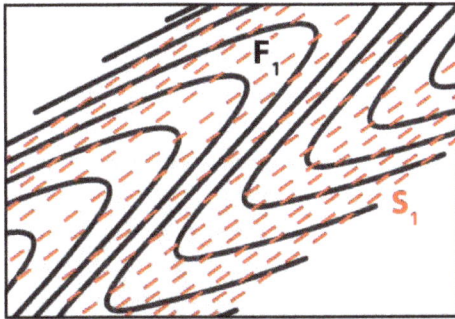

Fabric S_1 contemporary with folds F_1

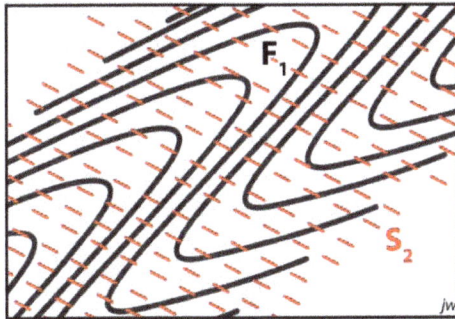

Fabric S_2 postdates folds F_1

—— Primary layering
– – – Tectonic fabric

Figure 6. Timing relationship between folds and fabrics.

The axial planar relationship is very useful when disentangling superimposed folds. It's conventional to number fold generations, starting with the earliest, as F1, F2, F3 etc. The corresponding axial planar foliations are called S1, S2 etc., where S1 is axial planar to F1, but is folded by F2. Sometimes the numbering scheme is even extended to bedding, which is labelled S0.

COMMON TYPES OF LINEATION, AND THEIR ORIGINS

Primary lineations: Primary lineations are less common than primary foliations. Typically, primary lineations in sedimentary and igneous rocks both reflect flow direction.

Mineral lineations: Mineral lineations are the linear equivalent of slaty cleavage and schistosity. Strain studies suggest that mineral lineations typically form parallel to the direction of maximum extension, the S1, or X axis of the strain ellipsoid. Like their planar counterparts. they probably form by a combination of grain rotation, solution, and grain growth during deformation.

Figure 7. Stretching lineation and flattening foliation, deformed conglomerate, Gansu China.

Stretching lineation: When a lineation is defined by domains that represent identifiable deformed objects in a rock, such as stretched pebbles in a deformed conglomerate, the fabric can be called a stretching lineations.

Figure 8. Crenulation lineation, BC.

Crenulation lineation: Crenulation cleavage is defined by closely spaced fold hinges. It's almost always a second or later generation of fabric: an initial deformation produces a foliation and later deformation folds that foliation to produce a crenulation lineation. Crenulation lineation is almost always visible when crenulation cleavage is present.

Figure 8. Mylonitic lineation, product of extreme shearing. Gansu China.

Mylonitic lineation: Extreme ductile shearing in shear zones tends to reduce the grain size of mineral grains, while stretching the original grains out into domains of extreme dimensions, and also producing strong crystallographic preferred orientation. This variety of stretching lineation is called **mylonitic** lineation. It's almost always accompanied by a foliation, producing an LS-fabric.

Intersection lineation: Whenever two intersecting foliations are present, there is always an **intersection lineation** defined by their intersection. In the field, although it's possible to calculate the intersection lineation from strike-and-dip measurements of the two foliations, it's usually much more accurate to measure the lineation directly.

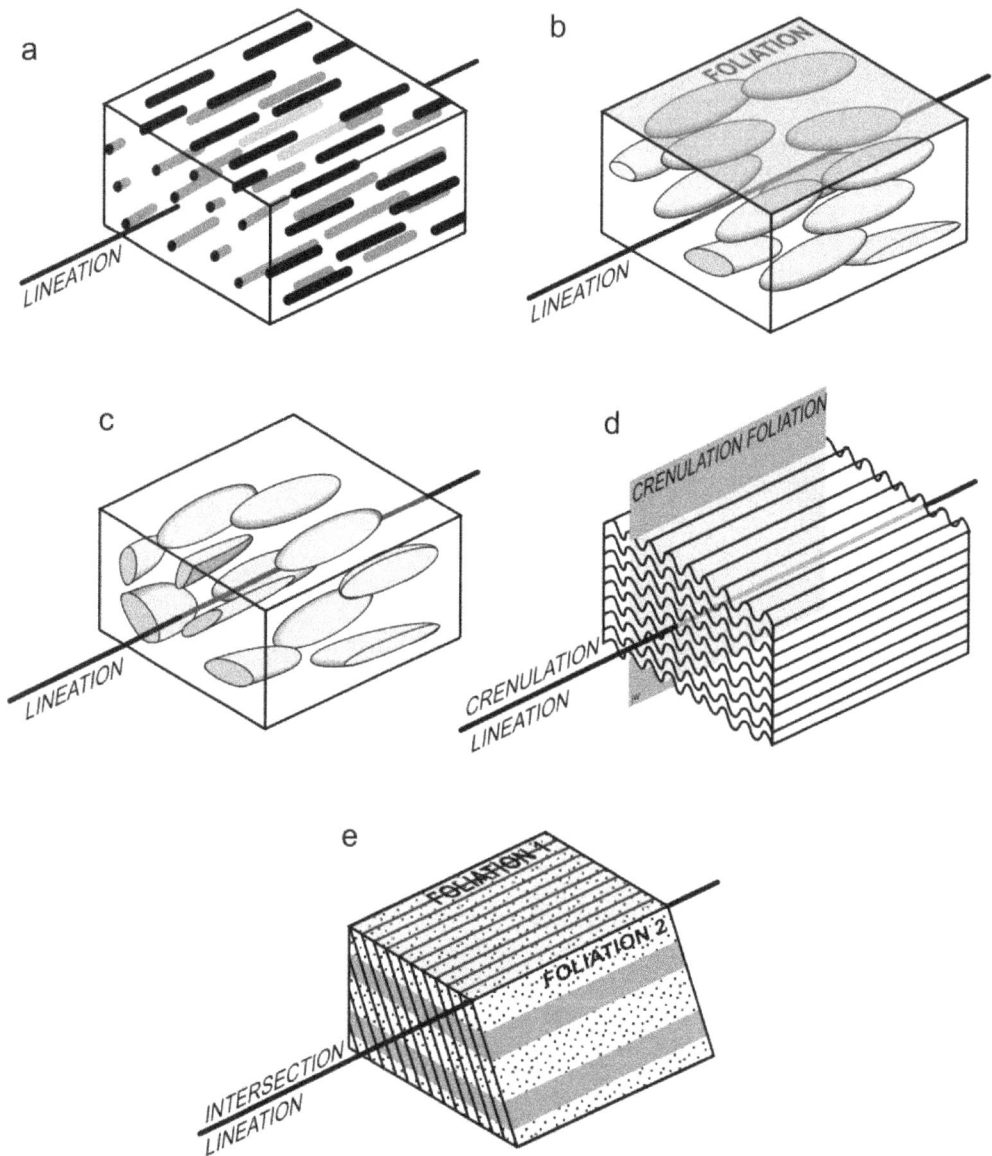

Figure 9. Different types of lineation and their fabric elements. The idealized ellipsoidal shapes in a – c could be individual mineral grains, clumps of metamorphic minerals, deformed pebbles, etc. (a) Simple lineation defined by elongate mineral grains or boundaries. (b) Combined lineation and foliation defined by bodies that are both elongated and tabular. (c) Bodies that are both linear and tabular are arranged so that only a lineation is defined. (d) Crenulation lineation and foliation defined by the hinges and axial surfaces of small folds or crenulations. (e) Intersection lineation defined by two differently-oriented foliations. Based on Hobbs, B.E., Means, W.D., and Williams, P.F. 1976. An Outline of Structural Geology. Wiley.

RELATIONSHIP OF LINEATION TO FOLDS

In folds exhibiting axial planar foliations, an intersecting lineation is generated where the cleavage planes intersect the folded surfaces.If the folds are sufficiently cylindrical, this intersection lineation is aligned with the fold axis.Measuring intersection lineation is an effective method for determining fold axis orientations in regions where fold hinges are inadequately revealed.Occasionally, a secondary lineation is seen in folded rocks exhibiting axial planar slaty cleavage.This lineation manifests as subtle streaks on the cleavage surfaces, oriented at a steep angle to the fold hinges.It is occasionally referred to as a downdip lineation due to its alignment in regions of gently descending folds.It likely originates as a stretching lineation and signifies the long axis of the strain ellipsoid (X). In higher-grade metamorphic rocks, stretching lineations, or mineral lineations believed to result from stretching, are occasionally observed parallel to fold hinges.This typically signifies that significant shearing has transpired, resulting in the formation of sheath folds.These will be examined in greater detail when we address mylonites later in the course.In areas with multiple generations of fabrics and folds, lineations are typically numbered L1, L2 etc., to correspond with fold and foliation generations.

LAB 6. INTRODUCTION TO FABRICS AND FOLDS

A fabric is fundamentally any structure that imparts varying qualities to a rock in one direction compared to another.Certain elements inside the rock exhibit a preferential orientation.The 'somethings' in this context refer to the fabric components.While a fabric may appear evident, a detailed examination may be required to identify its constituent pieces.Fabrics and folds, both resulting from strain, frequently function in conjunction.In the mapbased segment of this laboratory, you are required to elucidate the structure of a region characterized by both folds and textiles.Prior to commencement, ensure your familiarity with the preceding sections concerning fold overprinting and fabrics.Be advised that the samples may solely be accessible during laboratory hours.

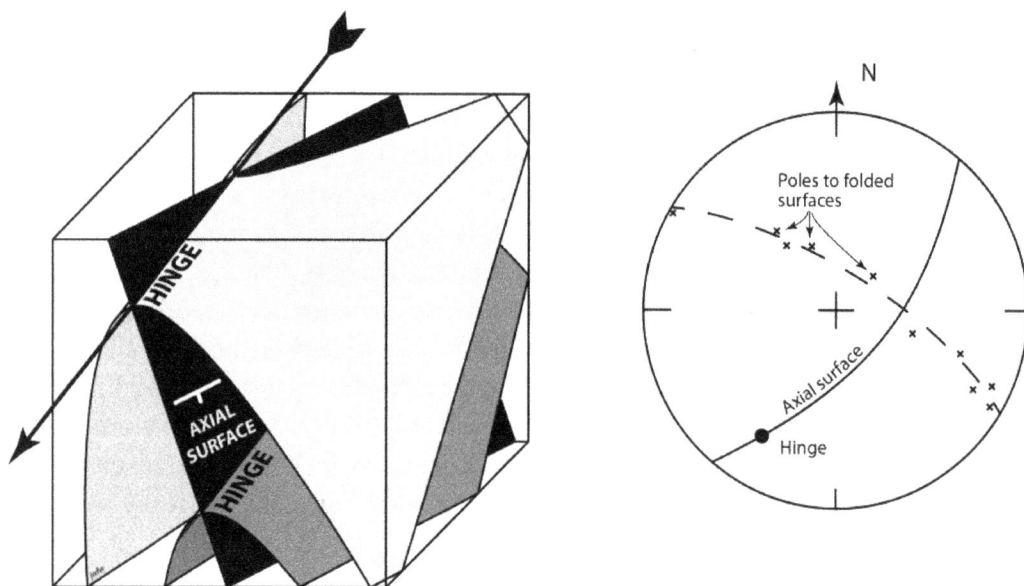

Figure 1. Relationship of fold hinge to axial surface. Left: block diagram. Right: stereographic projection.

Assignment.

1. *To illustrate the concept of fabric elements, you are provided with some everyday objects with simple fabric. For each fabric, determine:

⊙ Is the fabric a foliation or a lineation?

⊙ What are the fabric elements that define the fabric?

⊙ What shape are the fabric elements (tabular, acicular, some combination)?

2. *Look at one rock sample from each group. Note that some of the samples have more than one fabric. For each sample, record the following:

⊙ Identification of the sample;

⊙ How many differently oriented planar fabrics (foliations)?

⊙ How many differently oriented linear fabrics (lineations)?

For each fabric:

⊙ Describe the elements that define the fabric;

⊙ Does the fabric have a common name (slaty cleavage, crenulation lineation, etc.)?

⊙ What is the relationship of the fabric to other fabrics in the rock? Where possible, number the fabrics S1, S2, L1, L2, etc....

3. You are provided a map of a group of glacially smoothed, low-lying islands showing metamorphic rock outcrops (Map 1). The structures are typically of

the deeply eroded cores of many orogens. There are two generations of folds, with associated fabrics and asymmetric parasitic folds. A penetrative schistosity is present everywhere, together with a locally developed crenulation cleavage and crenulation lineation. In a few places, it is possible to see an intersection lineation between bedding and schistosity, though these two foliations are almost parallel in many outcrops, indicating that deformation has been intense.

a. Using an equal area projection, plot the orientations of the schistosity as poles.

b. Look at the folds on each island in turn. Notice that some of the folds deform the bedding but not the schistosity; these folds have axial planar schistosity, and therefore probably formed *at around the same time* as the schistosity. These folds are the first generation that can be identified in the area. Label them F1, in a distinctive colour (e.g. *red*). Other folds clearly formed *after* the schistosity because they fold the schistosity. Label these folds F2 in a contrasting colour (e.g. *blue*).

c. Label each F2 fold as S or Z based on its asymmetry. Use the same colour that you used to identify F2 folds.

d. Using your results from b and c, draw and label the principal *fold axial trace* for an F2 map-scale fold. Remember, a fold axial trace will separate S and Z senses of asymmetry. Label it using the F2 colour.

e. From your equal area projection, estimate the orientation of the F2 fold axis. Both the fold axis and the axial trace are lines that lie in the axial surface. Use these two lines to draw the axial surface as a great circle on your projection and determine its orientation. Add the crenulation cleavage and crenulation lineation to your projection. What is their relationship to the folds?

f. For each F1 fold, label it as S or Z based on its asymmetry, using the F1 colour. Draw the F1 axial traces on the map. (Remember, the F1 folds will probably have been refolded by F2, so you should not expect their traces to be straight!) Assuming the chlorite schist is the youngest unit, characterize the main F1 folds as either anticlines or synclines.

g. Add labels S0, S1, S2, L1, etc. to the legend where indicated by '.....'

h. Complete the map as far as possible in the unexposed areas between the islands to show the overall structure (to help you visualize the structure, you may wish to sketch a cross-section AB, but this is not required).

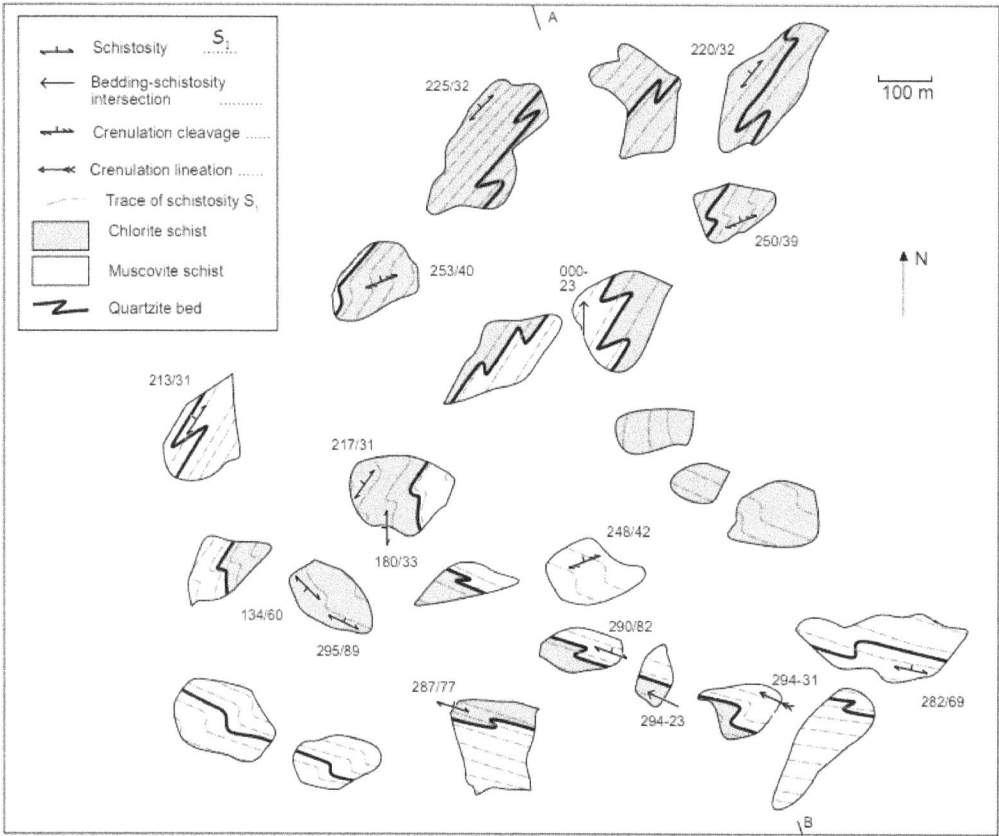

Lab 6 Map 1

9 Chapter | INTRODUCTION TO DYNAMIC ANALYSIS: STRESS

INTRODUCTION: AN OVERVIEW

Dynamics constitutes the segment of structural geology that pertains to energy, force, stress, and strength. It is crucial to differentiate dynamic notions from kinematic ones. Numerous errors have occurred in structural geology by individuals attempting dynamic analysis without first comprehending the principles of movement (kinematics). While the terms stress and strain may appear synonymous in colloquial usage, their scientific definitions diverge significantly. Stress is a dynamic concept, while strain is exclusively kinematic. In daily life, directly measuring force or stress is challenging. For instance, when you stand on a bathroom scale, you are distorting a spring (that constitutes strain!). The spring's well-defined dynamic features, where stress and strain are proportionate, enable us to deduce your weight from the observed strain. In geology, our understanding of the stress-strain relationship in rocks, particularly those subjected to deep burial and deformation over millions of years, is somewhat incomplete. Exercise caution while employing dynamic terminology; ensure you fully comprehend the subject matter.

FORCE AND STRESS

Units of force

Force is measure in Newtons where 1 N is the force necessary to accelerate a mass of 1 kg by 1 m/s^2.

Units of stress

In structural geology we are almost always interested in what a force does to some part of the Earth's crust, so we need a measure of *force concentration* or *force per unit area*. This is **stress**.

Note: some textbooks define two different quantities: **traction** is the force per unit area on a single plane, a **vector** quantity; **stress** is the total of forces acting on all possible planes that pass through a point in the Earth's crust, a **tensor** quantity. At this level we refer to both concepts as 'stress'; the sense is almost always clear from the context.

The unit of stress is 1 N/m^2 or 1 Pa (**Pascal**).

1 Pascal isn't enough to do detectable damage to any kind of rock. More useful units are:

1000 Pa = 1 kPa

10^6 Pa = 1 MPa

10^9 = 1 GPa

1 GPa is roughly the pressure at the base of the crust, about 30 km down.

An older unit of stress is the **bar**.

1 bar = 10^5 Pa or, more usefully 10 kbar = 1 GPa

There are other units out there. You may encounter the atmosphere (atm), and the pound per square inch (psi)

1 atm = 1.01 bar = 10100 Pa

1 psi = 690 Pa

All these units can be used to describe **pressure**. **Pressure** is the state of stress in a stationary fluid, like water. In fact, pressure is also known as **hydrostatic stress**. Hydrostatic stress is the type of stress experienced by a submerged submarine. Each 1 m^2 of the skin of the submarine experiences the same force, acting perpendicular to that surface.

Stress on a plane (a.k.a. traction)

In solids, the situation is more complex. Each surface of a mineral grain within the Earth experiences a different force concentration depending on its orientation. Also, some surfaces experience forces that are not perpendicular to the surface.

In fact, we can resolve the force per unit area (a vector) on a surface into two components.

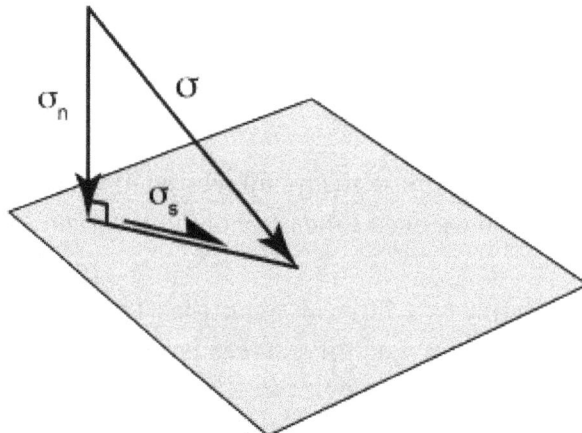

Figure 1. Stress, or traction, on a single plane.

Normal stress, also known as normal traction σ_n is the part of that stress that acts *perpendicular* to the surface. **Shear stress** also known as shear traction σ_s is the part that acts *parallel* to the surface.

Because most normal stresses within the Earth act inward, geologists represent compressive normal stress as positive, and use negative numbers for tensile stresses. When we start talking about stress and strain, this can cause confusion, because *positive stress (inwards) tends to cause negative extension (ie shortening)* and vice versa. (Engineers often use the opposite convention, which is mathematically more logical, but requires pressure to be a negative quantity, which is less intuitive for most people).

State of stress at a point

For any given orientation (strike and dip) of a surface passing through a given point in the crust, there is a different value of normal and shear stress. At first sight this is a bewildering mess of forces, all acting in different directions at the same point, but there are some relationships between the various forces that simplify things.

First, if we represent all the stresses acting on all the surfaces as vectors, drawn as arrows, the tails of those arrows make an ellipse (in 2-D) or an ellipsoid (in 3-D). The ellipsoid is called the **stress ellipsoid**.

Second, it's possible to prove that there are always three mutually perpendicular planes that experience *no shear stress*. These are **principal planes of stress**. The normal stresses they experience are the maximum, minimum, and an intermediate value of normal stress. We call them **principal stresses** and label them, in order:

$$\sigma_1 > \sigma_2 > \sigma_3$$

The directions of the principal stresses are called the **stress axes**.

You may have noticed that there are close analogies between stress and strain. Be careful not to confuse them! Stress is dynamic; strain is kinematic!

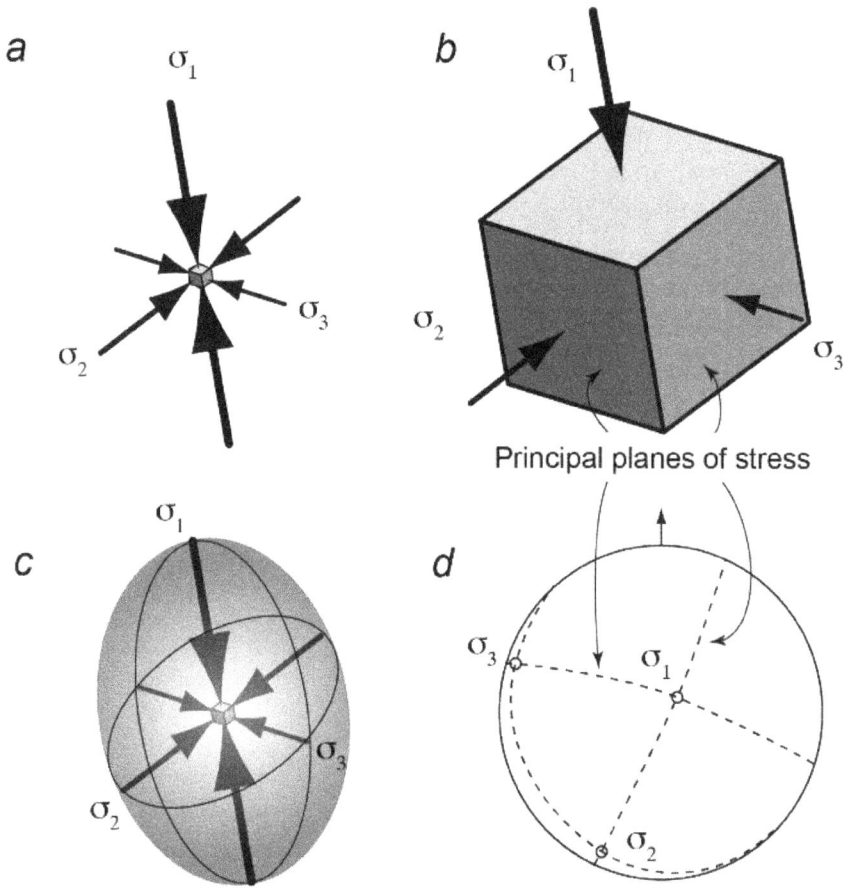

Figure 2. The state of stress at a point within the Earth can be represented in many different ways. (a) Stress axes and principal stresses. (b) Principal planes of stress; these planes experience zero shear stress. (c) Stress ellipsoid. (d) Stress axes and principal planes of stress on a stereographic projection.

Hydrostatic and lithostatic stress

Hydrostatic stress is the special case where $\sigma_1 = \sigma_2 = \sigma_3$ and is equivalent to 'pressure' in a fluid. Where the pressure is due to overlying rock, not fluid, the term **lithostatic stress** is sometimes used.

Non-hydrostatic and differential stress

In much of the Earth's crust, the state of stress is **non-hydrostatic**. However, pressure is still a useful concept. What we mean by pressure under those circumstances is **mean stress**. The mean stress is the average of the three principal stresses.

$$\sigma_m = (\sigma_1 > \sigma_2 > \sigma_3)/3$$

The mean stress is the part of the stress that *acts to change volume.* It's the most important type of stress for metamorphic petrologists, because high mean stress (high pressure) tends to produce dense minerals like garnet and glaucophane.

What about the rest of the stress? If we subtract the mean stress from each of the principal stresses, we get a "left over" stress system called the **deviatoric stress** defined by principal values.

$$\sigma_1 = \sigma_m, \quad \sigma_2 = \sigma_m, \quad \sigma_3 = \sigma_m$$

The deviatoric stress is the part of the stress that *acts to change shape,* and is the part of greatest interest to structural geologists.

A related concept is the **differential stress**. This is just the difference between the largest and the smallest principal stress: $\sigma_d = \sigma_1 - \sigma_3$.

Effective stress

A final concept to be aware of in dynamic analysis is that of **effective stress**. In porous rock, the pore spaces are typically filled with fluid, often water, but sometimes oil or natural gas. If that fluid is itself under pressure, it partially supports the mineral grains, and reduces the stresses between the solid parts of the rock, making them behave as if they were located at a shallower depth. The **effective stress** is the true stress minus the fluid pressure. It is the part of the stress that acts within the solid components of a porous rock.

Stress regimes

The Earth's surface is approximately a plane of zero shear stress (give or take a few ocean currents and wind storms). For this reason, near the Earth's surface, one of the principal stresses is approximately vertical, and the other two are approximately horizontal.

This idea was first promoted by Scottish geologist William Masson Anderson in 1905, and has become known as the Andersonian theory of stress. Anderson distinguished three near-surface tectonic regimes depending on which stress axis was vertical.

σ_1 vertical: **Gravity regime**

σ_2 vertical: **Wrench regime**

σ_3 vertical: **Thrust regime**

In a general way, these three regimes correspond respectively to typical states of stress near the three types of plate boundary: spreading centres, transform faults, and subduction zones.

STRESS-STRAIN RELATIONSHIPS

Experimental vs. geological strain rates

Experimental rigs are used to study the types of stress that are necessary to produce different kinds of strain in rocks. A great deal has been learned from such experiments. However, it's important to realize one major limitation: time. Geological strain rates are of the order of 10^{-12} – 10^{-15} strains per second. In the lab, if we don't want to run our experiments for hundreds of years, it's not feasible to achieve strain rates much below 10^{-8} strains per second. To get geologically meaningful results we have to *extrapolate* experiments to much slower strain rates.

Elastic

When rocks are subjected to small strains at low confining pressure (or mean stress) we find that the stress is proportional to the strain, and the strain is recoverable (i.e. it goes away when the stress is removed).

This type of stress-strain relationship is called **elastic**. The elastic behaviour or rocks allows them to store strain energy and to transmit seismic waves.

Brittle

As stress and strain are increased, eventually most rocks undergo a catastrophic loss of strength, with the release of stored strain energy. In an experiment this is called **brittle fracture** or **brittle failure**. In an experimental rig the result is a loud bang and the sample disintegrates. In the Earth's crust the result is an **earthquake.**

Plastic

If the confining pressure is higher still, or the temperature is raised, a different type of behaviour occurs. After an initial phase of elastic deformation, the sample starts to deform in a **ductile** manner: it flows without breaking. This deformation is also non-recoverable, but it occurs without complete loss of strength. The sample shortens in the σ_1 direction, and thickens parallel to σ_3.

In **ideal plastic** behaviour, a sample shows no deformation at all until a certain stress (yield stress) is reached. Thereafter, it deforms freely so that however much shortening is imposed by the rig, it's impossible to get the stress to go any higher.

Real rocks are a bit more complicated than the ideal. They typically show some elastic deformation below the yield stress, and with continuing plastic deformation the stress may rise a little or fall a little.

Viscous

At temperatures near their melting point, certain rocks exhibit a more straightforward flow behavior devoid of yield stress. In viscous behavior, even a minimal application of tension will initiate strain. The strain rate is directly proportional to the stress. Viscous behavior is occasionally referred to as Newtonian. Water, air, magma, and rock salt may exhibit approximately Newtonian behavior.

Competence

What factors determine a rock's response to stress? A multitude of variables governs the particular response in a certain context. Confining stress, or mean stress, which is approximately synonymous with 'pressure,' is one such variable; mean stress escalates with depth into the Earth. Temperature significantly affects rock strength and generally rises with depth in accordance with the geothermal gradient. The strain rate, defined as the velocity of rock deformation, constitutes a third variable. The rock composition significantly influences the outcome as well. Each of the aforementioned idealized behavioral categories is associated with a metric that quantifies a rock's strength or resistance to stress. Young's modulus pertains to elastic deformation; the differential stress at failure characterizes brittle behavior; yield stress defines plastic deformation; and viscosity describes viscous behavior. Actual rocks exhibit complex combinations of different behavioral patterns, rendering these numbers exceedingly difficult to quantify. Nevertheless, in the field, we can frequently identify rock types that have experienced varying degrees of strain, positioned adjacent to one another inside the same outcrop, suggesting they have likely endured comparable stress histories. In these circumstances, it is advantageous to possess a broad phrase for resilience to stress. The term is competency. If we note that slate layers have experienced significantly more strain than the interbedded quartzite layers, we may infer that the quartzite exhibited greater competence.

10 Chapter

INTRODUCTION TO FRACTURES

INTRODUCTION: AN OVERVIEW

Figure 1. Joints in limestone are often pathways for groundwater flow. Here the groundwater has dissolved the limestone surface to produce an irregular topography. Port au Port Peninsula, Newfoundland (boot for scale).

The majority of the structures examined in this course are the result of ductile deformation, which is promoted by elevated temperatures, significant confining stress, minimal strain rates, and reduced rock strength. Conversely, fractures signify the brittle collapse of rocks due to applied differential stress. Fracturing is generally favored by low pressure or mean stress, low temperatures (when rocks exhibit maximum strength), high strain rates, and robust or competent rocks, including well-cemented sandstones, limestones, and igneous intrusive rocks. Fractures are flat or slightly curved failure surfaces resulting from the brittle failure of rocks. An extension fracture is defined as a fracture when the rock masses on either side have experienced modest separation. If the two rock masses have moved laterally relative to one another, the cracks are classified as shear fractures. In the context of natural fractures, extension fractures are typically referred to as joints in the field. Shear fractures are classified as faults when the rocks on one side have been substantially displaced along the fracture surface. This part analyzes joints; subsequent sections address flaws. Joints and veins within the Earth's crust hold significant value for humanity. Joints represent zones of vulnerability in both natural and artificial formations; failure along these surfaces has resulted in devastating landslides and rockfalls. In the subsurface, joints serve as conduits for the migration of fluids, such as water, oil, and natural gas. Comprehension of fluid movement across fractures is essential for groundwater utilization and the extraction of oil and natural gas. The interval between rock masses along a joint surface may become saturated with minerals deposited from groundwater, resulting in the joint being termed a vein. Numerous economically significant minerals are extracted from veins.

GEOMETRY OF JOINTS AND VEINS

The orientation of a fracture, similar to any planar surface, is described by its strike and dip. The position and direction can be depicted by structural contours at map size. The rocks just above a fracture form the hanging wall, while those directly below provide the footwall (Figure 2). Consequently, the plane inclines towards the hanging wall by definition. Clearly, these phrases are irrelevant if the fracture is vertical, in which scenario the walls are more accurately designated as 'north wall' or 'southeast wall', etc., based on the strike.

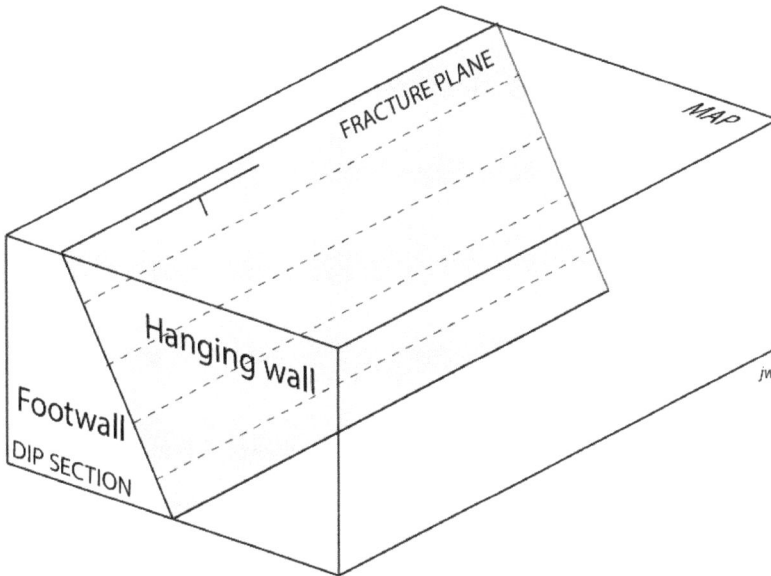

Figure 2. Block diagram of a dipping fracture, illustrating the hanging wall and the footwall.

The distance between the two walls of a fracture is referred to as the fracture aperture.Certain fractures have a consistent aperture across extensive distances, although others display some degree of fluctuation.Fractures do not propagate indefinitely through the Earth's crust.The termination point of a fracture trace is referred to as a fracture tip.In three dimensions, the apex of a fracture is a linear characteristic referred to as a tip line.

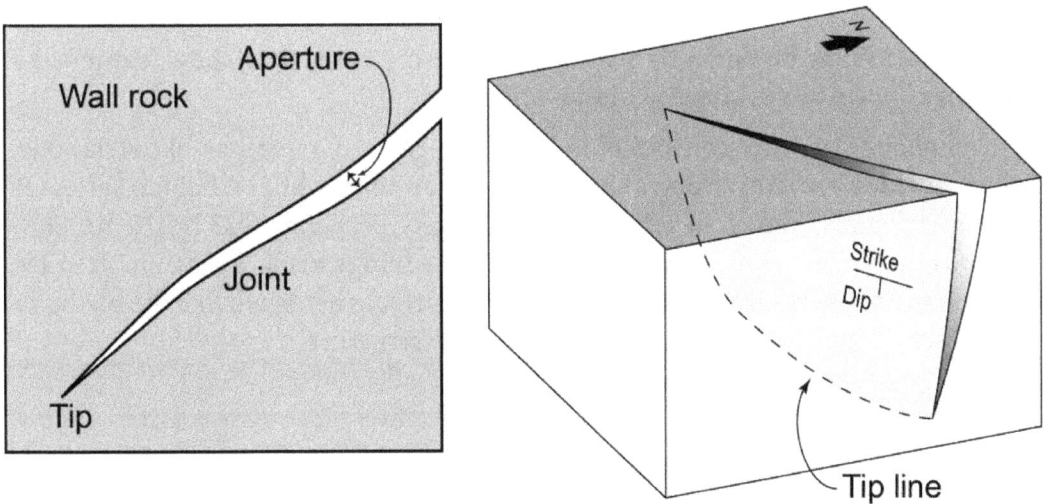

Figure 3. Geometry of a joint. Left: in plan view. Right: in 3-D.

Joints are very common in many outcrops, but their orientations are typically not random. It is common to observe many joints in approximately the same orientation. These are known as a **joint set.**

A combination of joint sets, cross-cutting each other in a regular way, is known as a **joint system.**

FEATURES OF JOINT SURFACES

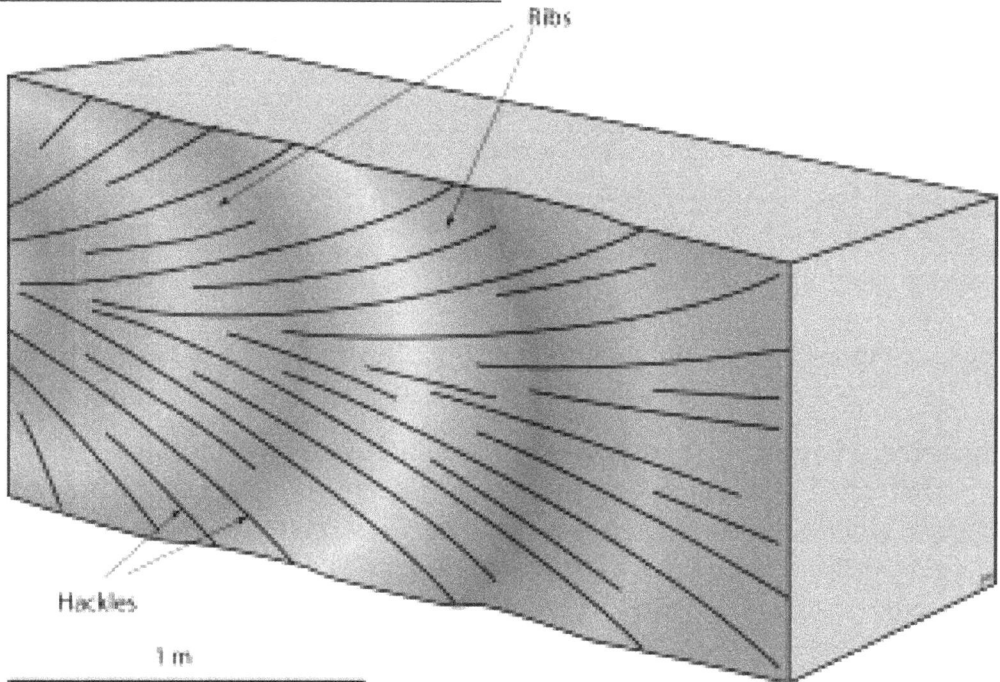

Figure 4. Geometry of plume structures on a joint surface.

Joint surfaces may be perfectly planar, but in many cases they show feather-like markings known as **plumose** or **plume structure.**

Typical plume structure consists of two parts. A pattern of feather-like striations, called **hackles** radiates from a central point or line, while a bolder series of concentric bands, known as **ribs** is approximately perpendicular to the hackles. Towards the tip of a joint, in what is known as the **fringe area,** it is common to see that the surface splits into a number of smaller surfaces arranged like the blades of a fan. Each of the blades may have its own plume structure.

Figure 5. Plumose markings including hackle marks, Blairmore Group, Bow River Valley, Alberta.

Figure 6. Plumose markings: hackles and ribs, Blairmore Group, Bow River Valley, Alberta.

Plume structure can be simulated in laboratory experiments. It records the process of fracture **propagation**, and the dynamics of the fracture tip.

Fracture propagation is the process whereby a fracture grows within a rock, and involves the outward migration of the fracture tip. Fracture propagation may be very fast: speeds up to those typical of seismic waves (a few kilometres per second) are possible.

Because the fracture tip at any moment is the boundary between unfractured rock and rock that has been weakened by fracturing, the orientations of the stress axes are modified in the vicinity of the tip. This in turn causes the variations in fracture orientation as the fracture propagates.

The kinematics of fracture propagation may be interpreted from plume structure. Hackles diverge in the direction of propagation, while concentric ribs mark successive pauses in the position of the fracture tip.

FEATURES OF VEIN FILLS

Two contrasting types of vein fills are common.

When a joint aperture has been held open by fluid pressure during hydrothermal activity, mineral crystals may be deposited on the walls, producing a vein. Crystals that grow into the fluid-filled spaces may be **euhedral** (showing clear crystal surfaces) but more typically the crystals run into one another producing **compromise boundaries** that are also approximately planar. There is often a tendency for crystals to get larger towards the centre of a vein.

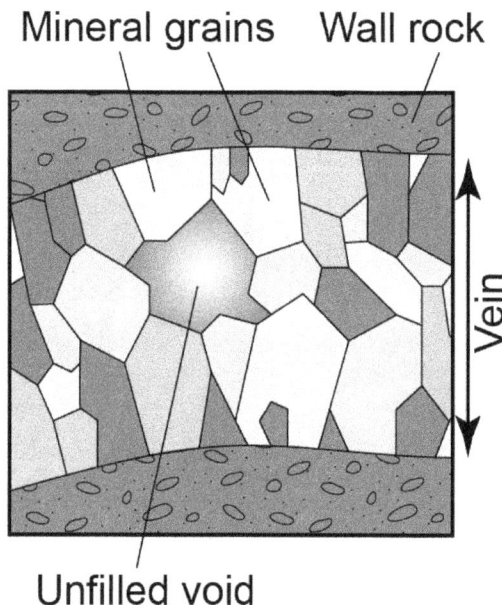

Figure 7. Vein with non-fibrous fill and remaining unfilled void space

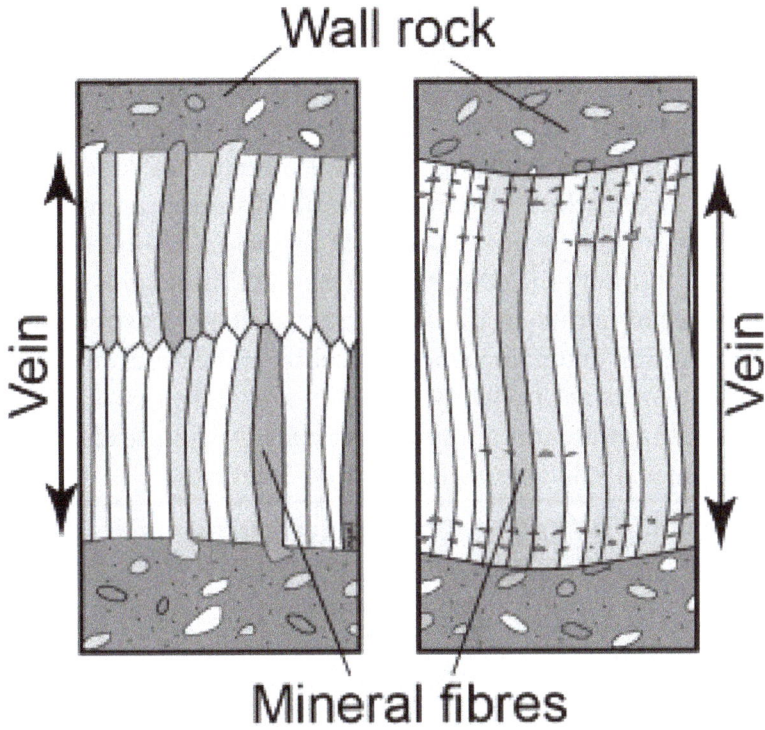

Figure 8. Fibre-filled veins, illustrating features of syntaxial (left) and antitaxial (right) fibre fills.

Figure 9. Fibrous quartz filling a vein. The quartz fibres grew simultaneously with vein opening, and track the relative movement of the two walls. SE Ireland.

Conversely, when a joint is filled with minerals deposited during the opening of the fracture, fibrous fills may occur. A narrow fissure is generally filled by an initial layer of crystals. This is succeeded by a reopened opening and the deposition of further mineral material, and so on. This procedure is referred to as cracksealing. A series of fibers intersects the vein. The fibers monitor the kinematic process of venous dilation. Two distinct forms of fiber fillings are present in veins. When repeated breaking transpires at the center of the vein, the fibers are generally in crystallographic continuity with the grains in the wall. These veins are characterized as syntaxial. In certain cross-sections of syntaxial veins, it is feasible to identify the line of demarcation extending along the center. If the repeated cracking occurs at the edge, the vein is antitaxial. Antitaxial veins often lack a singular central line; nonetheless, they may contain several strands of wall-rock inclusions aligned parallel to the vein walls, indicating consecutive fractures.

DYNAMIC INTERPRETATION OF FRACTURES

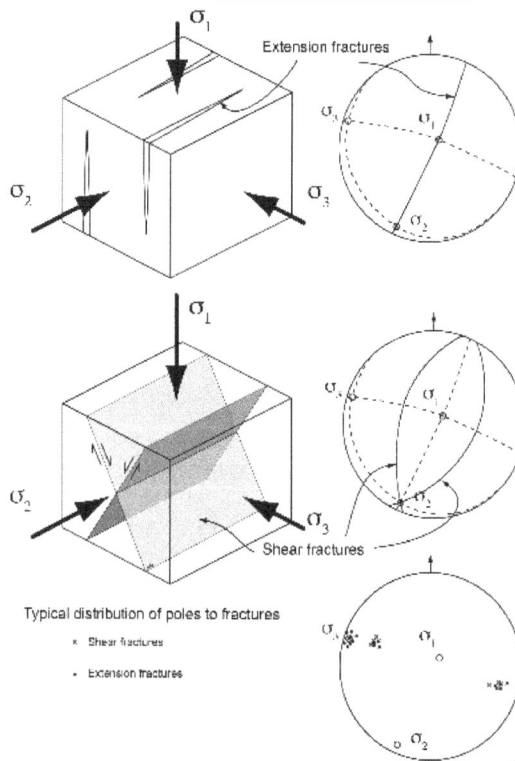

Figure 10. Typical relationships between stress and fractures in block diagram and stereographic projection. These examples show fractures formed in the gravity regime (σ1 vertical), but the same angular relationships between the structures and stresses apply in other stress regimes. Top: extension fractures. Middle: conjugate shear fractures. Bottom: distribution of poles to extension and shear fractures on a stereographic projection.

There are some characteristic relationships between the orientation of fractures and the stress axes that give us useful information.

Extension fractures

Rocks fractured under low mean stress (or 'confining pressure') are likely to break along fractures perpendicular to σ3.Upon examining the fractures, we ascertain that they are extension fractures, characterized by the separation of the two sides.

Conjugate shear fractures

At higher pressure, failure typically occurs along two families of planes, breaking rocks into wedge-shaped fragments. The angle between the two families is about 60°. The planes intersect each other in a line parallel to σ_2. The maximum principal stress σ_1 bisects the acute angle between the planes, and the minimum stress σ_3 bisects the obtuse angle.

If we examine the fractures, we find that they are **shear fractures**: the two sides of the fracture have slid past each other. The sense of shear is such that the acute angles at the edges of the fragments have been pushed inward. Movement is inwards along the σ1 direction and outward along σ_3.

COMMON TYPES OF JOINT AND VEIN SYSTEMS

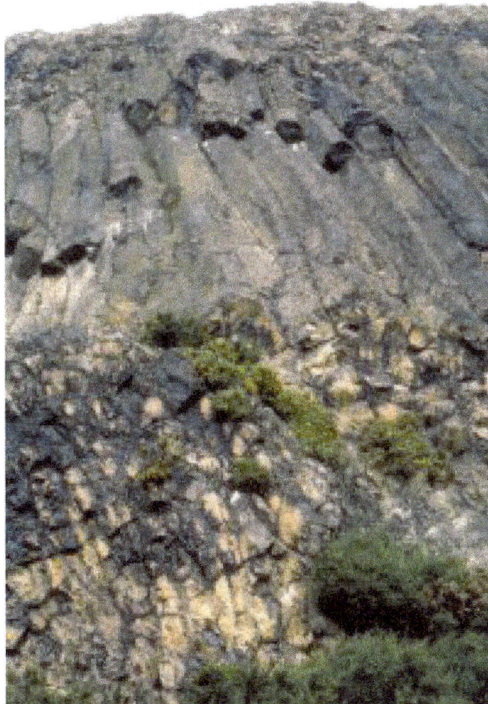

Figure 11. Columnar joints in sill. Salisbury Crags, Edinburgh, Scotland.

Primary joints in igneous rocks

The primary joints in igneous rocks generally arise from contraction following cooling from magmatic temperatures.Most of these joints develop perpendicular to the surfaces of lava flows or the interfaces of intrusions.In tabular igneous formations such lava flows, sills, and dykes, the joints can exhibit a columnar structure and display a very uniform arrangement of polygons, predominantly hexagons, in crosssection.Research indicates that a hexagonal configuration lowers the energy required to create new surfaces for a specified degree of contraction.

Joints in the host rock of intrusions

Joints are prevalent beyond igneous intrusions. Two prevalent forms are radial joints and concentric joints. These joints often arise from extension either during the vigorous intrusion of magma or during the subsequent differential cooling of the intrusion.

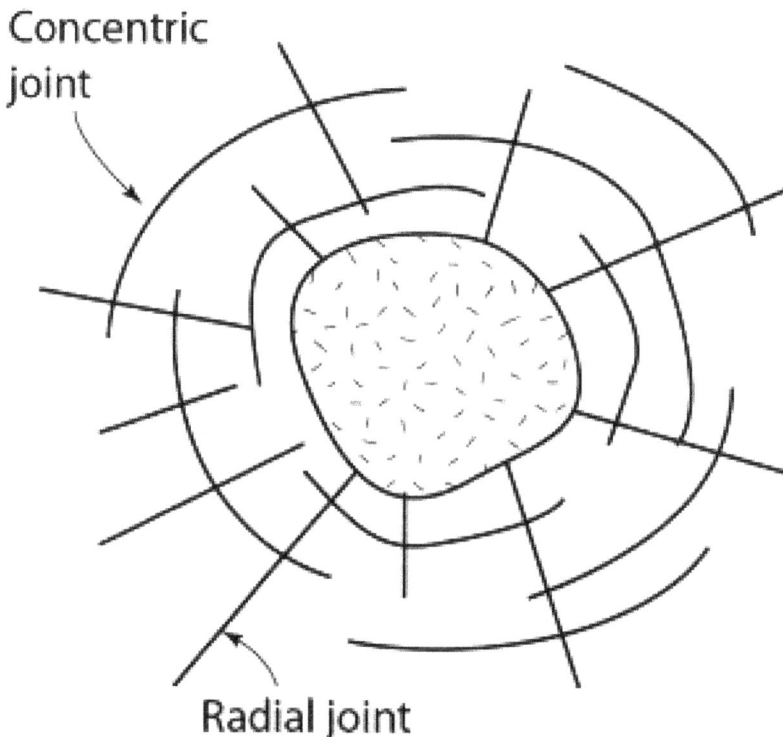

Figure 12. Radial and concentric joints around an intrusion.

Joints related to erosion and exhumation

Joint sets frequently develop parallel to the topographic surface in regions that have seen fast erosion. These are referred to as sheeting or sheet joints, and the process is termed exfoliation. These joints are particularly prominent in intrusive igneous rocks such as granite, which are generally isotropic. In extreme instances, granite exhibiting numerous sheet joints may be mistaken for a bedded sedimentary rock when observed from afar.

There are several possible causes of exfoliation.

- In regions where the crust is in horizontal compression, σ_3 is vertical. As overburden is removed by erosion, and the mean stress is reduced, σ_3 becomes negative (tensile). The rock tends to expand upward and does so by forming roughly horizontal joints.

- Secondly, residual stresses may be locked into rocks at the time of formation, particularly during the cooling of intrusions. If tensile stresses are perpendicular to contacts, these may lead to the formation of joints parallel to the contacts when erosion reduces the overall pressure.

- Finally, weathering of igneous rocks leads to significant volume changes, as water is absorbed, and clay minerals form. These volume changes may set up stresses that encourage exfoliation behaviour.

Joints related to faults and shear zones

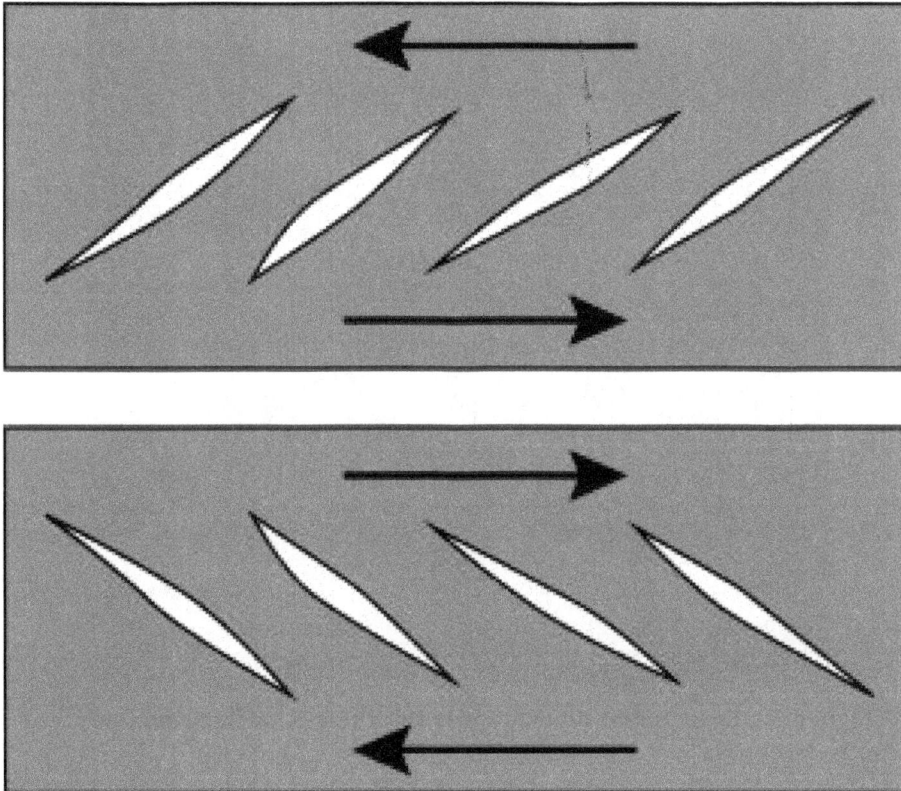

Figure 13. En-echelon joints and veins related to sinistral shear (top) and dextral shear (bottom).

Faults and shear zones are typically associated with high densities of joints and other fractures. One common geometry reflects the distribution of stress in a developing fault zone. Repeated joints or veins are oriented at an angle of about 45° to a developing fault in a configuration called *en echelon* (derived from a diagonal military formation). The joints or veins are formed by extension, and are oriented perpendicular to σ_3.

Figure 14. En-echelon quartz veins in schist. Amaliapolis, Greece.

Joints related to folds

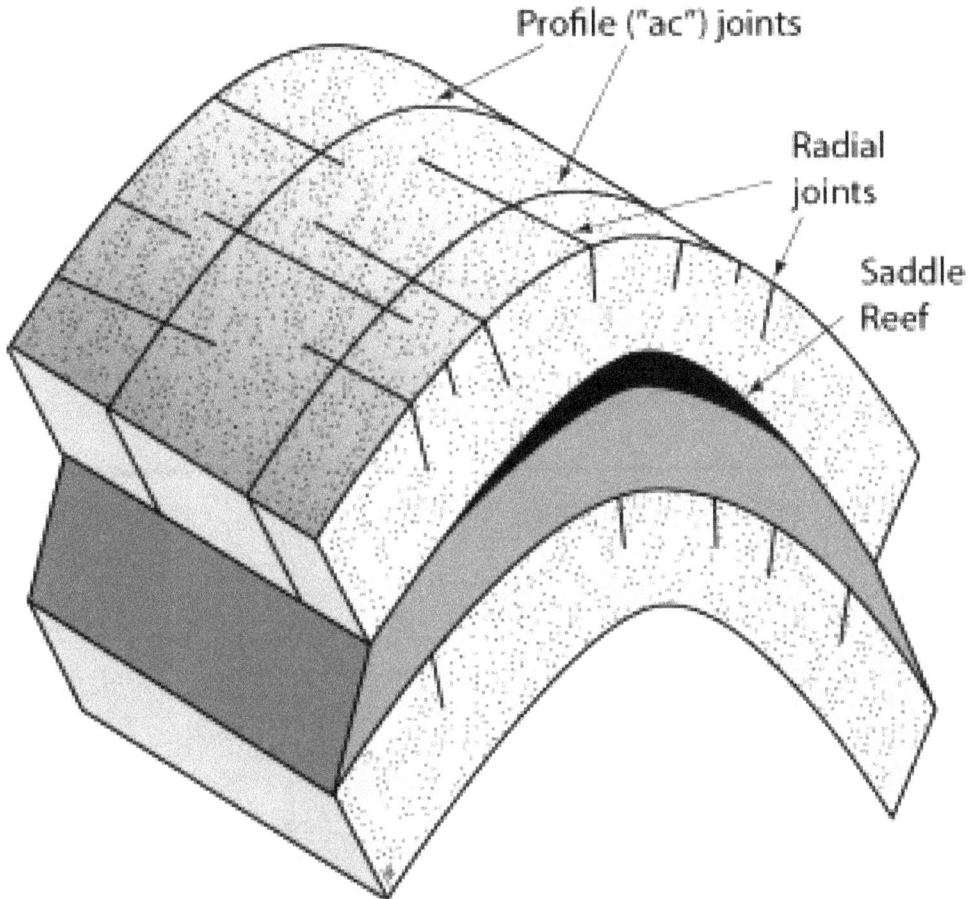

Figure 15. Common configurations of joints and veins related to folds.

Complex joint patterns may be observed in folded layered rocks, especially sedimentary rocks that were near the brittleductile transition during folding. The most prevalent kind are radial joints that run parallel to fold hinges and are approximately perpendicular to the layers. They predominantly occur on the external regions of buckled competent layers and likely develop as a reaction to the stretching of the outer arc of the layer during folding. A secondary set of joints is occasionally observed parallel to the folded strata, notably at the inner surfaces of buckling competent layers. These likely occur when the incompetent layers lack sufficient ductility to occupy the gaps in the hinges between competent layers. Veins in this location have been significant sources of gold in several goldfields, especially in Australia and Nova Scotia, where they are referred to as saddle reef veins. A third category of fold-associated veins manifests parallel to the profile plane of folds. These are logically referred to as profile veins, but they are frequently termed ac-veins, derived from an antiquated nomenclature for folds that involves three axes: a, b, and c. The fold axis, as currently designated, was referred to as axis b in this arrangement. The profile of joints and veins suggests that an extension component occurred along the length of the fold hinges.

LAB 7. FRACTURES

This lab introduces two novel approaches in stereographic projection that are very beneficial for analyzing joints and veins. Both pertain to diminutive circles.

Orientation of veins from multiple boreholes

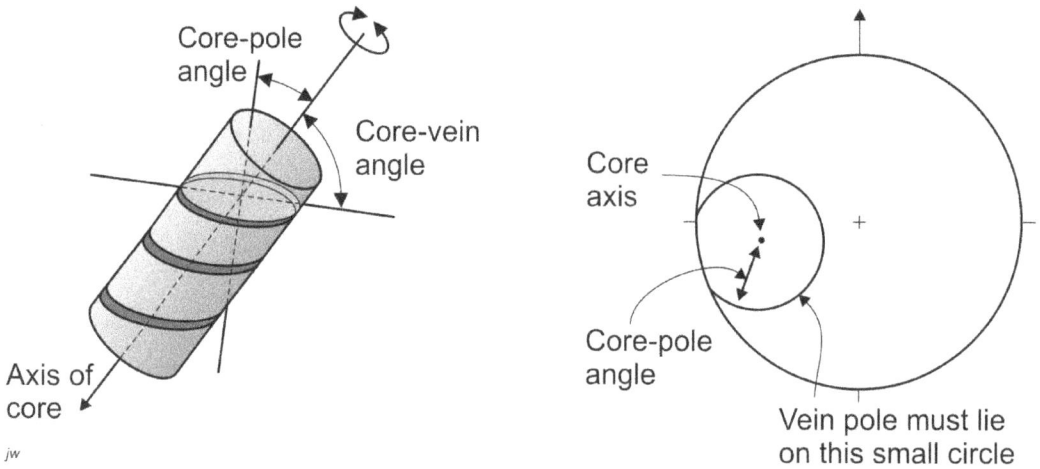

Figure 1. Un-oriented core showing orientations of veins (left) and locus of possible vein poles (right).

Many exploration programs are carried out to determine the location and orientation of buried mineralized veins. Typically, diamond-drill holes will be drilled at various angles into the ground and cores will be recovered, showing the orientation of the veins. Unfortunately, the drilling operation does not allow core to be recovered without rotating it in the drill hole, so a single core does not provide a unique indication of the strike and dip of a vein. All we can determine from a single core is that the pole to a vein must lie on a **small circle** on the stereographic projection (Fig. 1). It turns out that intersecting small circles from 3 cores are necessary to uniquely determine the orientaiton.

Constructing small circles can be complicated. If the cores are horizontal, it is easy, because the Wulff and Schmidt nets have small circles centred on the primitive. However, if the cores are plunging (the more usual case), constructing small circles is more involved. There are two methods. Method 1 depends on a property of the

Wulff net, on which all small circles plot as true circles (as shown in Fig. 1) so it is possible to construct them with a plotting compass; however, the centre of the projected circle does not match the axis of the small circle in 3-D, so the centre of the point has to be found first. The process is simple if the small circles do not cross the primitive, but a little more involved if they do, as shown in Figure 2. Method 2 depends on **rotating** the orientations of the two cores to the primitive to construct the small circles there. Once the intersections are found, the rotation is reversed to find the true orientation of the intersections. The second procedure works on either the Wulff or Schmidt net and is explained in detail in Figure 3.

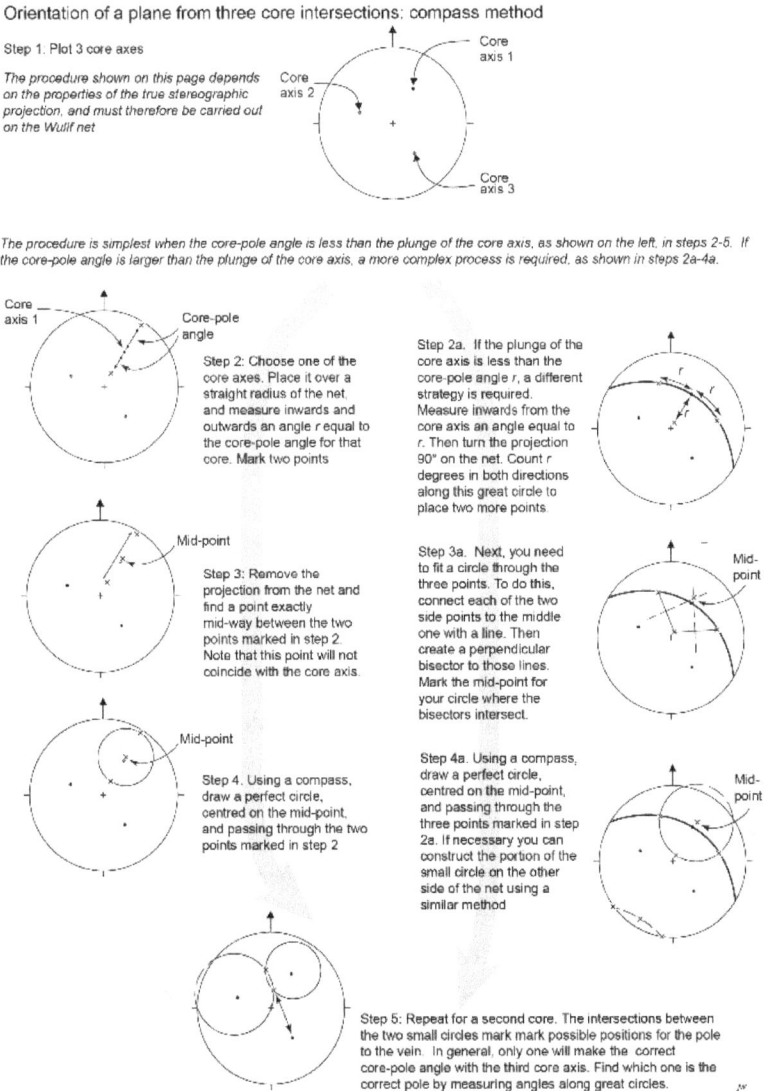

Orientation of a plane from three core intersections: compass method

Step 1: Plot 3 core axes

The procedure shown on this page depends on the properties of the true stereographic projection, and must therefore be carried out on the Wulff net

Core axis 1

Core axis 2

Core axis 3

The procedure is simplest when the core-pole angle is less than the plunge of the core axis, as shown on the left, in steps 2-5. If the core-pole angle is larger than the plunge of the core axis, a more complex process is required, as shown in steps 2a-4a.

Core axis 1

Core-pole angle

Step 2: Choose one of the core axes. Place it over a straight radius of the net, and measure inwards and outwards an angle r equal to the core-pole angle for that core. Mark two points

Step 2a. If the plunge of the core axis is less than the core-pole angle r, a different strategy is required. Measure inwards from the core axis an angle equal to r. Then turn the projection 90° on the net. Count r degrees in both directions along this great circle to place two more points.

r

Mid-point

Step 3: Remove the projection from the net and find a point exactly mid-way between the two points marked in step 2. Note that this point will not coincide with the core axis.

Step 3a. Next, you need to fit a circle through the three points. To do this, connect each of the two side points to the middle one with a line. Then create a perpendicular bisector to those lines. Mark the mid-point for your circle where the bisectors intersect.

Mid-point

Mid-point

Step 4. Using a compass, draw a perfect circle, centred on the mid-point, and passing through the two points marked in step 2

Step 4a. Using a compass, draw a perfect circle, centred on the mid-point, and passing through the three points marked in step 2a. If necessary you can construct the portion of the small circle on the other side of the net using a similar method

Mid-point

Step 5: Repeat for a second core. The intersections between the two small circles mark mark possible positions for the pole to the vein. In general, only one will make the correct core-pole angle with the third core axis. Find which one is the correct pole by measuring angles along great circles.

Figure 2. Method 1 for orienting a planar surface intersected by three differently-oriented drill holes.

Orientation of a plane from three core intersections: rotation method

Step 1: Plot 3 core axes

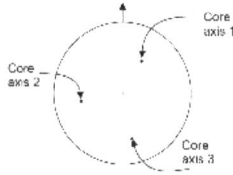

Step 2: Choose 2 axes and draw a
great circle through them.

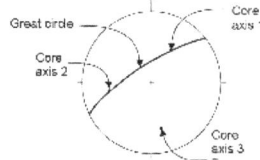

Step 3: Note the dip d of the great circle. This is the angle
through which the core axes will have to be rotated to
place them on the primitive. Note the strike of the great
circle. This will be the rotation axis

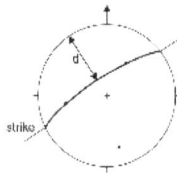

Step 4: 'Rotate' all the core axes until the great circle
is horizontal and two of the rotated core axes lie on
the primitive. All 3 core axes move d degrees along
small circles.

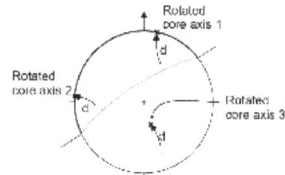

Step 5: Using the net, draw small circles, centred on
the two core axes that are on the primitive. The
radius of each small circle is the core-pole angle for
that core.

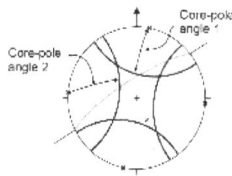

Step 6: The intersections between the small circles
mark possible positions for the pole to the vein. In
general, only one will make the correct core-pole
angle with the third core axis. Find which one is the
correct pole by measuring angles along great circles

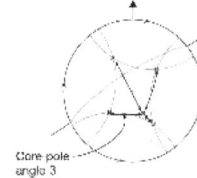

Step 7: 'Rotate' the correct pole back to its true
location by reversing the process of step 3, and
convert to the strike and dip of the plane.

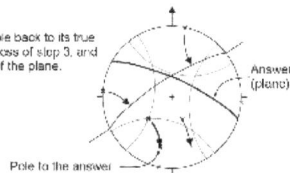

Figure 3. Method 2 for orienting a planar surface using three differently-oriented drill holes.

Density plots of poles

Occasionally, numerous joints are assessed in a specific region.Geologists may seek to ascertain the average orientation of the joints or to articulate the variability in their orientation.Conventional arithmetic techniques for computing averages are ineffective for orientation data, as orientations are vector values.To comprehend the issue, attempt to calculate the average of two trends near north, 359° and 001°. The arithmetic mean of 180° or due south is evidently nonsensical.Commencing with a contoured plot illustrating the density of poles on a stereographic projection is typically more advantageous.Due to the analysis of pole densities, it is essential to employ an equalarea projection (Schmidt net) for this procedure.

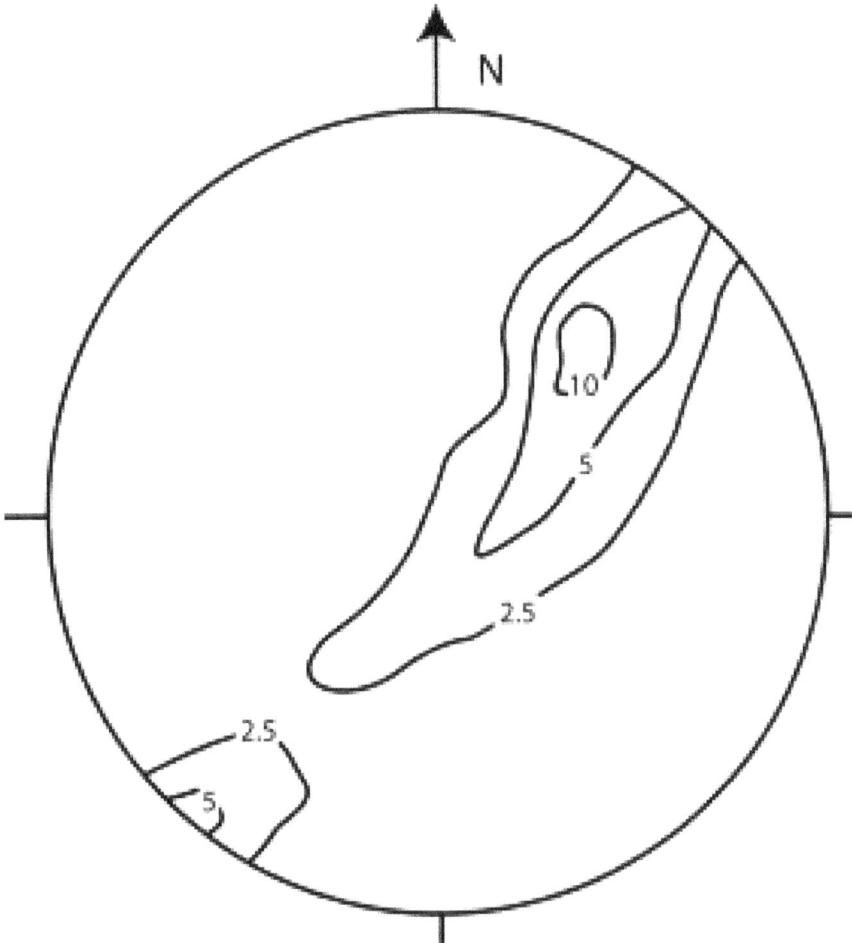

Figure 4. Example of a contoured equal area projection (contours show density of data per 1% area of projection).

The first step in any contouring process is to superimpose the data on a grid. For this purpose you can either tape the projection down onto a piece of graph paper or you can draw a grid directly on the projection. For the example in the assignment, a grid has been drawn on a plot of poles.

Next, we need to find how many data points lie in each 1% of the net. To do this, a device called a counting circle is helpful. It can be a piece of paper with a circular hole, or a piece of tracing paper with a circle drawn on it.

In either case, the diameter of the circle must be exactly one tenth of the diameter of the equal-area net; its area will then be 1% of the net area.

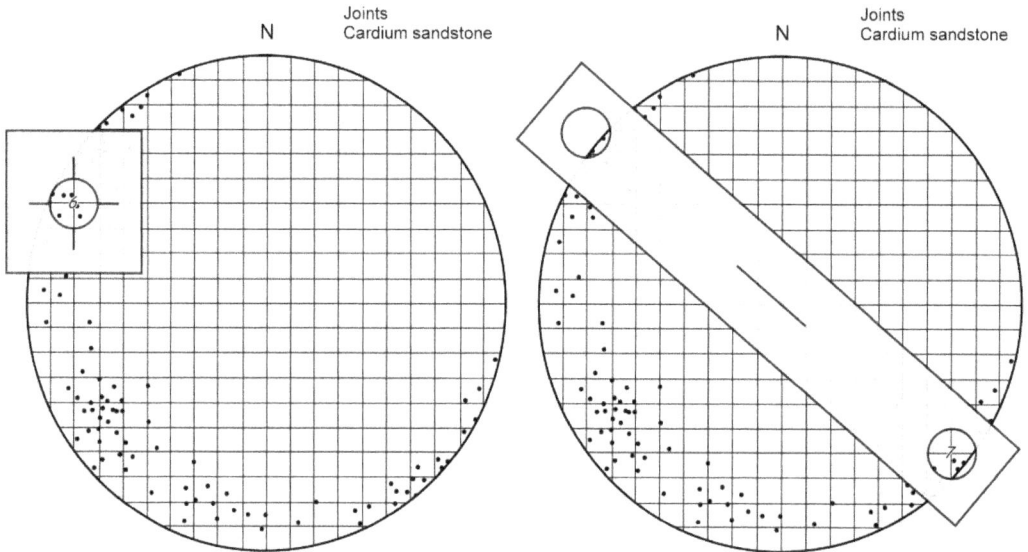

Figure 5. Procedure for counting pole density using a counting circle (left) and a peripheral counter (right).

Place the counting circle directly over a grid intersection on the equal area plot. Count the number of poles that fall within the circle. Write this number on the grid intersection. Repeat with all the intersections in the plot.

For intersections that fall near the edge of the plot, part of the counting circle will fall outside the primitive. The points that 'should' be in this part of the counting circle will appear on the opposite side of the net. We therefore use a second device, called a peripheral counter, which has two counting circles, their centres spaced apart by the same distance as the diameter of the net. A slot in the centre of the peripheral counter fits over the thumb tack in the middle of the plot, so as to keep the counter correctly aligned. When one hole is centred over a grid intersection near the primitive, the circle on the opposite side contains the extra points that should be counted.

Finally, make a smooth contour map of the results. As in all contouring, make sure that higher numbers all appear on one side of a contour, and lower numbers on the other. Contours should never branch. In contouring, remember that contour lines that disappear over the primitive should reappear on the opposite side of the net. Number the contour lines in percent, meaning "percent of the data occurring in each 1 % of the plot area".

Assignment

1. A mineral exploration company drilled three diamond drill holes to intersect a vein set in the subsurface. The angles between the axes of the cores and the veins are shown in the following table.

Drill hole #	Trend & plunge of hole	Core-vein angle	Core-pole angle
1	340-70 NW	40	50
2	080-76 NE	65	25
3	210-68 SW	54	36

Calculate the strike and dip of the vein set.

2. You are provided with an equal area projection of 100 poles to conjugate **shear fractures** in the Cadomin Formation from the Alberta foothills. To find the maximum densities of points, it is helpful to construct a contoured plot. A grid has been constructed across the projection to assist with contouring.

a) First, cut out the centre counter and peripheral counting circles precisely. The counting circles have a diameter exactly 10% of the plot, which means that their area covers 1% of the plot. Use the counting circles to mark densities at the each grid intersection. Contour the resulting densities on the grid. Remember the principles of contouring that you used in the first lab: each contour should separate higher from lower values; make the contours as smooth as possible while honouring the data; contours should never branch. In addition, remember you are really contouring on the surface of a sphere, and the pattern is symmetrical on the lower and upper hemispheres. Therefore, if a contour disappears over the primitive, it should reappear on the opposite side of the projection. When you are done, make a clean copy with selected contours on a separate tracing sheet.

b) Mark estimated centres, or modes, of the two density clusters. These are *poles* to the typical conjugate shear planes. From these poles, determine the strike and dip of the two planes, and draw the planes as great circles on your plot.

The maximum principal stress is predicted to bisect the acute angle between the two planes. The minimum principal stress bisects the obtuse angle, and the intermediate stress is parallel to their intersection.

c) Mark the principal stresses on your contoured projection and determine the plunge and trend of each.

d) An oil company is interested in flow through extension joints in the Brazeau formation. Predict the orientation of these joints based on your estimates of the directions of the principal stresses.

Lab 7. Question 2. Plot of poles to 100 shear fractures in Cardium Sandstone, Alberta Foothills. [PDF]

Lab 7. Question 2. Counting Tool [PDF]

3. *Look at the geological map of Canmore (east half) which shows numerous

faults. Notice the band of dark blue Palliser Formation in the region north of Exshaw.

a) Without using the cross-section, what is the evidence that the Palliser Formation dips to the SW in this area?

b) Now look at the Palliser Formation on cross-section 1, which extends into the sheet Canmore (west half). Place two sheets of tracing paper so as to cover the cross-section, and trace only the base of the Palliser Formation (DPa) and the faults that offset it. For each of the *named* thrust faults, answer the following questions:

c) Does the fault place older over younger or younger over older strata?

d) Estimate the the dip separation of the base of the Palliser formation. Note – if the fault is curved, you can approximate the separation with a series of straight-line measurements with a ruler.

e) Estimate the throw and the heave. (These are straight line measurements even if the fault is curved.)

4. *Strictly speaking, you have no evidence for the direction of movement on the faults: – it could be exactly parallel to the cross-section or there could be a component of strike-slip movement, in and out of the page. However, it is clear that there has been substantial overall shortening of the rocks in the hanging wall of the lowest, McConnell Thrust. Estimate this shortening in the line of the cross-section, as follows:

a) Measure the total original length l_0 of the segments of the base of the Palliser Formation in the cross-section.

b) Measure the present-day distance l between the easternmost outcrop of the base of the Palliser, and the western edge of the section.

c) Calculate the shortening in kilometres: $l_0 - l$.

d) Calculate the longitudinal strain as a value of **stretch:** $s = l / l_0$.

e) Calculate the longitudinal strain as a fractional change in length or **extension:** $e = (l - l_0) / l$

11
Chapter

INTRODUCTION TO FAULTS

INTRODUCTION: AN OVERVIEW

Fractures are classified as faults when there is considerable displacement of one side in relation to the other, parallel to the fracture plane. Faults have significantly influenced the economic aspects of natural resource exploration. Faults influence fluid dynamics within the Earth's crust, hence regulating the distribution of water, oil, and natural gas. Fractured material along a fault line may create a porous breccia.Fluids across breccia can deposit precious minerals.Faults are significant to people as they induce earthquakes.A comprehensive lexicon has emerged concerning defects, their geometry, and kinematics.It is essential to differentiate between descriptive (geometric) terminology, which conveys the orientation of a fault and the displacement of layers on each side, and kinematic terminology, which delineates the distance and direction of fault movement.

FAULTS, FAULT ZONES, AND SHEAR ZONES

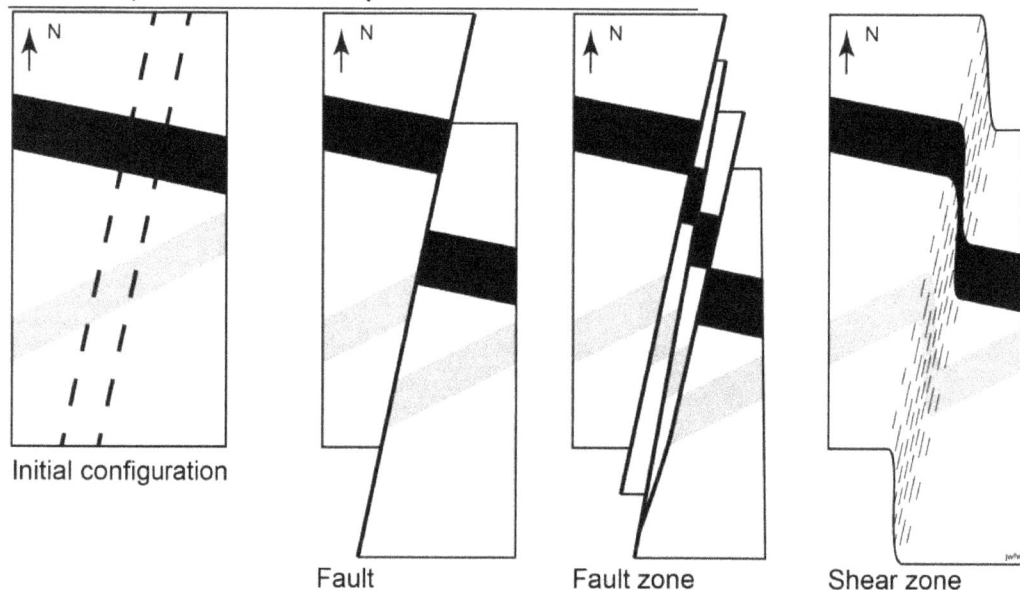

Figure 1. Fault, fault zone, shear zone.

Strictly speaking, a **fault** is a single fracture surface. However, many mapped faults turn out to have multiple fault strands, all roughly parallel, but branching and joining ("anastomosing") along their strike. This type of fault array is called a **fault zone**. The total offset of the fault zone is distributed across the zone. Each fault in the zone offsets the rocks on either side by a small amount. These add up across the fault zone to a much larger offset.

A related structure is a **shear zone.** A shear zone is the ductile equivalent of a fault zone – a belt of rock across which movement has caused a significant amount of offset between the two sides. However, a shear zone is a **ductile** structure, typically formed at depth where the deformation of the mineral grains is plastic, not brittle. The movement in a shear zone is distributed across the zone, rather than being restricted to discrete brittle faults. It is likely that most faults that deform the upper crust pass at depth into shear zones in the lower crust.

At [pb_glossary id="779"]map scale[/pb_glossary], faults and shear zones look the same – a line across which older structures are offset. At outcrop (mesoscopic) or microscopic scale they look rather different.

The orientation of natural faults often varies and they eventually disappear along strike; however, they do tend to have surfaces that are more planar than other types of geological surfaces, at least over distances of a few kilometres. Faults can be recognized at a variety of scales from centimetre-scale offsets in an individual outcrop to faults that can be traced on the ground for tens to hundreds of kilometres such as the San Andreas Fault.

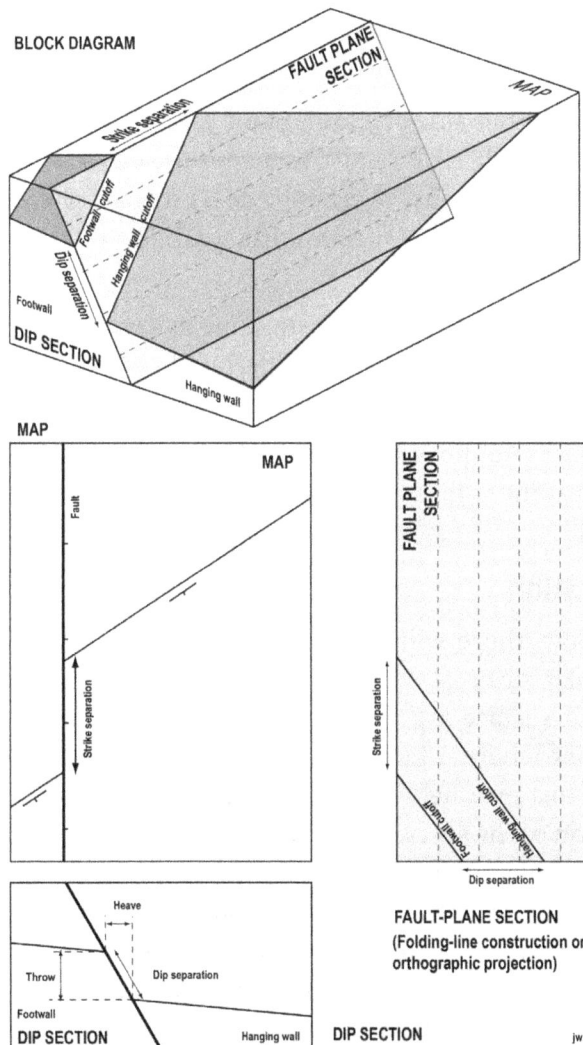

Figure 2. Block diagram of a simple fault offsetting a single surface (dark grey). Below, the same fault is shown as it might appear on a map and in a cross-section parallel to the dip-direction (a dip section). The grey diagram shows a map of the fault plane, known as a fault plane section.

FAULT GEOMETRY

Strike and dip

A fault is a planar geologic structure. Like any planar structure, it has an orientation that may be characterized by **strike** and **dip.** For small faults, it may be possible to walk up to an outcrop and measure the orientation with a clinometer. Large faults tend to be poorly exposed, because rocks close to the fault plane are fractured and broken, and therefore are easily weathered. For large faults, the orientation may be more easily determined by drawing structure contours. If the fault is planar, the strike and dip will be constant, and the structure contours will be parallel, straight, and equally spaced. If the fault is curved, then structure contours may show changes in orientation and spacing.

The blocks of rock on either side of a fault plane are the walls of the fault. If a fault has a dip (other than 90°) then one wall overhangs the other. For example, if a fault dips east, then the east wall must overhang the west wall. The wall located on the down-dip side of the fault is called the **hanging wall.** The wall located to the up-dip side of the fault is called the **footwall**.

If these are not clear, think of a geologist standing in a small cave exactly on a fault plane that dips moderately. The geologist's *feet* will rest naturally on the *footwall*, while the *hanging wall* will *overhang* the geologist's head.

If a fault is vertical, it is not possible to characterize a footwall and a hanging wall. Under those circumstances it is better to use compass directions to identify the walls: e.g. "east wall".

Offset and separation

If a fracture is a fault, there will also be **offsets** of beds or other older surfaces that are cut. On a map or cross-section, the points where the trace of a fault cuts the trace of an older surface are called **cutoff points** (or cut-off points). In three dimensions, the lines where a fault intersects an older surface are called **cutoff lines** (Fig. 3). The distance between the two cutoff lines, for a given surface, is called the **separation** of the surface. In plan view, as seen on a map of a horizontal surface, the distance of offset measured parallel to the strike of the fault is called the **strike separation.** The strike separation can be **sinistral** (also known as **left-lateral**) or it can be **dextral** (also known as **right-lateral).** In cross-section view, parallel to the dip of the fault, the separation is called **dip separation.** If the beds in the hanging wall are offset below those in the footwall the separation is **normal.** If the beds in the hanging wall are offset above those in the footwall then the separation is **reverse.** The vertical and horizontal components of dip separation were very important in old mining operations, and are known as **throw** and **heave** respectively.

It's important to realize that these measurements of separation are geometric. They tell you little about kinematics. In the diagram below, the arrows on the fault plane show that an infinite number of slip directions is compatible with a given fault separation.

One feature that may be recognized in the field and is a common result of relatively recent fault activity is the development of a **fault scarp**, which is that part of the failure surface exposed by movement of the rock masses. In studies of recent faults **(neotectonics)** the fault scarp may give an indication of separation.

However, over time the high-standing block tends to be levelled by erosion. The surface expression of the scarp is subdued or eliminated and may be of little help in locating ancient faults. Where fault scarps do occur along ancient faults, the direction of slope tends to be determined by which side has the more erosion-resistant rocks, not by the sense of separation on the fault. In fact, locating faults in Precambrian terrains may be quite difficult because rocks affected by faulting tend to be easily eroded. Such faults often occur in low-lying regions occupied by swamps, streams or heavy vegetation.

FAULT KINEMATICS: MEASURING SLIP

Slip

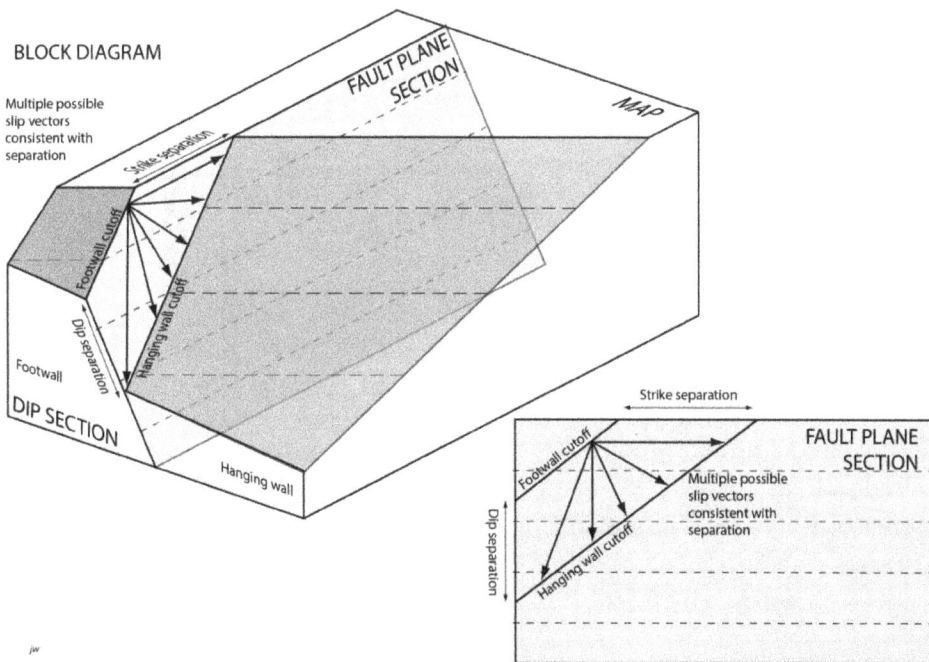

Figure 3. Fault plane intersecting a bedding surface, producing cutoff lines in the hanging wall and footwall. Multiple possible slip vectors are compatible with the same geometry.

The **slip** is a line that lies in the fault plane; slip represents the distance and direction of movement between the two blocks of rock on either side of the fault. The slip is a **vector** – it has a distance and direction. The direction of slip may be specified by **trend and plunge**, or by **rake** within the fault plane. When the rake of the slip is close to the *strike* of the fault, the fault is called a **strike-slip fault**. When the rake is near 90°, close to the *dip* of the fault, the fault is a **dip slip fault**.

Figure 4. Slickenlines on a fault surface, Arisaig, Nova Scotia.

It is important to understand that slip cannot be determined from the separation of a single offset surface; we need additional information such as **slickenlines** (scratches or fibres on the fault surface) to give us the direction of the slip. Figure 3 illustrates the problem: multiple slip vectors would be consistent with the separations shown by the fault.

Piercing points

If we have a line that is cut and displaced by the fault, we can solve uniquely for the slip vector. The intersection of a line with a fault plane produces a point, called a piercing point. If we can locate the piercing points for the blocks on either side of the fault then we will be able to determine the displacement vector as shown in Figure 5.

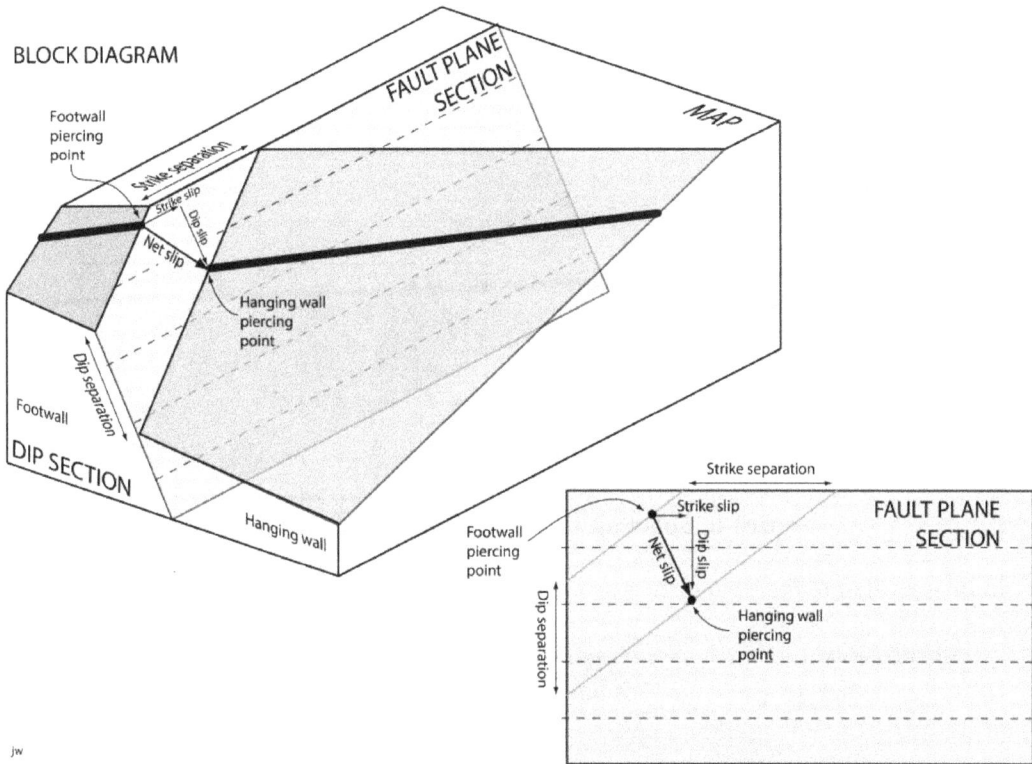

Figure 5. Piercing points on a linear feature uniquely define slip.

The **slip** is *a line that lies in the fault plane*, that represents the distance and direction of movement between the two blocks of rock on either side of the fault. The slip is a **vector** – it has a **distance** and **direction**. The direction of slip may be specified by **trend and plunge**, or by **rake** within the fault plane. When the rake of the slip is close to the *strike* of the fault, the fault is called a **strike-slip fault**. When the rake is near 90°, close to the *dip* of the fault, the fault is a **dip-slip fault**.

The question is, what will we use for a line cut by the fault? Recall that two non-parallel planes intersect along a line, and you have your clue. If two planar features (such as veins, unconformities, formation boundaries, or igneous dykes) intersect each other, then their line of intersection may in turn pierce the fault plane. Figure 6 shows an example where an igneous intrusion is truncated by an unconformity at a subcrop line. Displacement along the fault offsets the subcrop line producing piercing points for hanging wall and footwall. A unique vector joins the piercing points, characterizing the fault displacement.

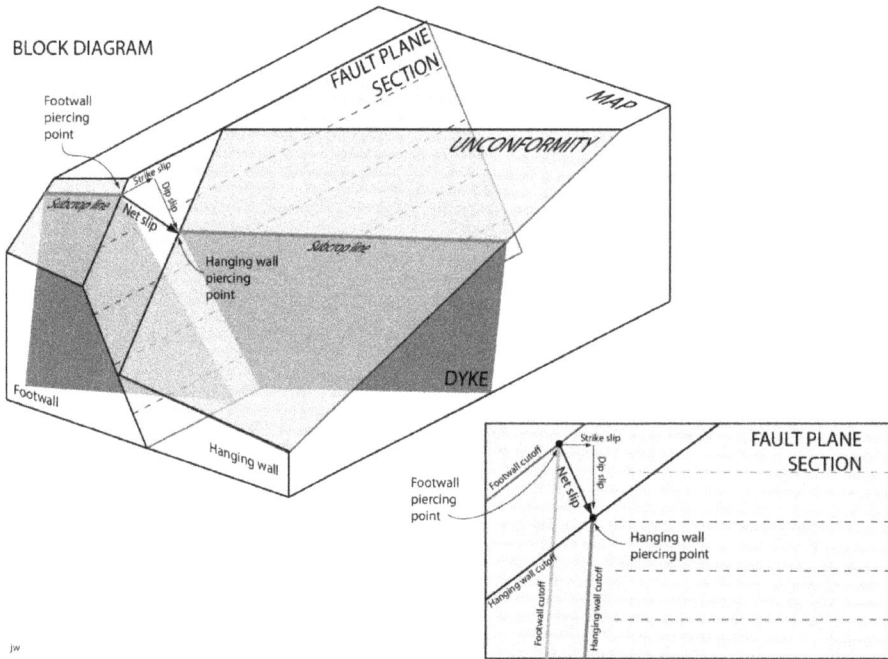

Figure 6. Subcrop line defining slip of a fault.

Fold hinges also make excellent piercing points. Figure 7 shows an example.

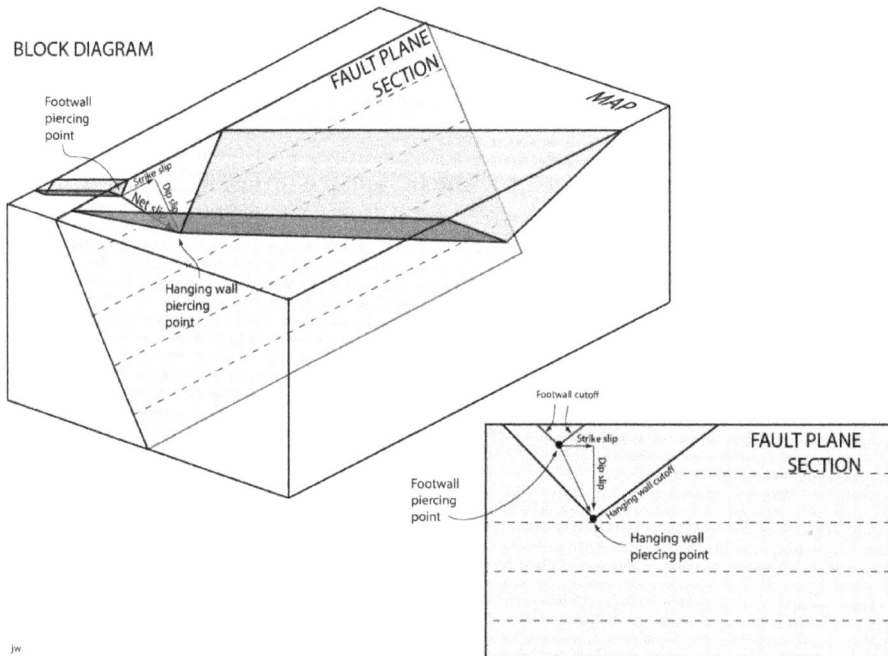

Figure 7. Fold defining slip of a fault.

Figure 8 shows a second circumstance in which the net slip can be calculated. In this example, striations on the fault surface (known as slickenlines) allow the direction of slip to be determined. It is then possible to construct artificial piercing points on cutoff lines at opposite ends of a single slickenline. The amount of slip can be measured between the constructed piercing points.

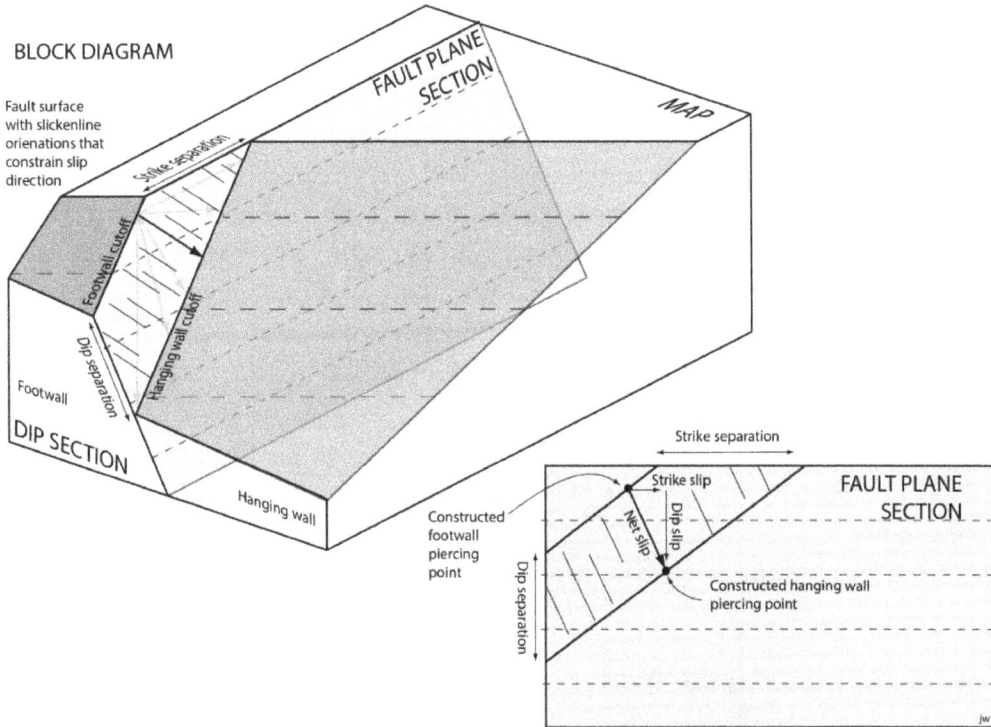

Figure 8 Slickenlines defining slip of a fault.

Components of slip

Once the net slip vector is known, it can be broken down into components. The component of slip parallel to the strike of the fault is the **strike slip.** The strike slip may be characterized as **sinistral** (or **left-lateral**) or **dextral (right-lateral).** The component parallel to the dip of the fault is called the **dip slip.** The dip slip may be characterized as **normal** or **reverse,** depending on whether the hanging wall moved up or down relative to the footwall. If the dip slip is much larger than the strike slip, then the fault is a **dip-slip fault;** the rake of the slip in the fault plane is close to 90°. If the strike slip is much greater, then the fault is a **strike-slip fault;** the rake of the slip is close to zero or 180°. If the dip slip and strike slip are of comparable magnitudes, then the fault is an **oblique-slip fault.**

Fault-plane sections

Most calculations involving fault slip are best carried out on a cross-section that lies in the fault plane. Figures 3, 5, 6, and 7 are each illustrated with a diagram showing a fault-plane cross-section. Most of the significant measurements of separation and slip can be made directly on the fault-plane cross-section. Of course, in general such a cross-section will not be vertical. Special techniques are necessary for the construction of non-vertical cross-sections. A plan view of a dipping geological surface is sometimes called an **orthographic projection** (because features are projected perpendicular to the surface onto a plane sheet of paper) and sometimes called a **folding line construction** (because we treat the sloping fault plane as if we have folded it around one of its structure contours until it is in the horizontal plane of a table top.) Typically, we begin by placing a sheet of paper with its edge along one of the structure contours on the fault plane. This line becomes the **folding line**. Next, we use a stereographic projection or contours to determine the **rake** of various lines on the fault surface and draw them in.

Sometimes it also helps to draw the projected structure contours as horizontal lines on a fault-plane section, to show elevations above sea level (see fault-plane sections in Figures 3, 5, 6, and 7). The lines are not spaced as they would be on a vertical cross section. If the contour interval is i and the fault plane has dip d, the spacing of the contours on the fault plane section will be:

$$i' = i \, / \sin(d)$$

Features of faults in the field

In the description of individual faults, it is generally possible to distinguish a **core zone** from a surrounding **damage zone.**

In the **core**, the rock has been broken up and moved around by fault movement so that the original pieces are separated from their neighbours. In the core zone it is no longer possible to see how the original pieces for faulted rock were fitted together.

In the **damage zone**, the rocks are fractured and may show other deformation features such as folds. However, the movement is not so great as to obliterate earlier structures, and it is at least possible to see how the pieces of damaged rock can be fitted back together.

FAULT ROCKS: FAULT CORE

Figure 8a. Fault breccia, Guysborough County, Nova Scotia

Several names are given to fragmented and other rocks of the fault core.

Breccia is a rock composed of typically angular fragments of the walls of the fault that have been rotated and moved out of their original positions. The fragments in a fault breccia may be huge, or may range down to about 2 mm. The term **microbreccia** may be used for a fine-grained breccia. Although the grain-size limits for breccia and microbreccia are not precisely standardized, it is reasonable to use microbreccia for fragments that are 2-4 mm in diameter (by analogy with the grain size of microconglomerate in sedimentary petrology).

Cataclasite is fragmented material that is sand-size or smaller. Note that neither breccia nor cataclasite typically has a fabric – the fragments are randomly oriented.

Figure 9. Microscope thin section of cataclasite. Field of view ~1 cm.

Gouge is the name given to clay-rich fault material, typically produced from faulting of fine-grained sediments. Because clay can behave in a ductile manner even under near-surface conditions, gouge can develop some fabric, because the clay particles become smeared into an orientation parallel to the fault plane. Gouge is typically very soft in outcrop. In the subsurface, gouge is very important, as a thick gouge layer may make a fault plane impermeable to fluids, whereas breccia and cataclasite are typically permeable.

[Note on **mylonite:** Mylonite is a name given to fine-grained rock that has formed through ductile shearing and recrystallization in a **shear zone.** Mylonite is distinguished from breccia and cataclasite because it has a strong metamorphic fabric and often shows spectacular crystallographic preferred orientation – CPO – when viewed under the microscope in thin section. Confusion arises because when mylonite was first described in the 19th century, it was thought that the fine grain size resulted from **brittle** fracturing – the word 'mylonite' is derived from a Greek word for milling flour. In the 20th century the study of metals in industrial processes showed that **ductile** deformation could lead to the grain-size reduction seen in mylonites]

Pseudotachylite is material that was melted by heating during fault movement. Pseudotachylite is typically dark and very fine-grained or glassy. It may occur in small dykelets that penetrate the wall rock of the fault. Pseudotachylite takes its name from an old term for a very fluid igneous lava, tachylite. *Pseudo*-tachylite looked the same but had a very different origin.

DEFORMATION ADJACENT TO FAULTS

Riedel shears are small shears that develop in response to stresses in the fault walls during fault propagation and movement. **Synthetic Riedel shears,** or **R-shears,** have an orientation at about 15° to the main fault plane, and display the same sense of shear as the main fault, as shown in Fig. 10. Less commonly, **antithetic Riedel shears,** or **R'-shears,** are also developed, at about 75° to the fault plane. These have a sense of shear opposite to that of the main fault. The synthetic and antithetic shears form a conjugate set, and therefore can be used to indicate the orientation of the stress axes when they formed.

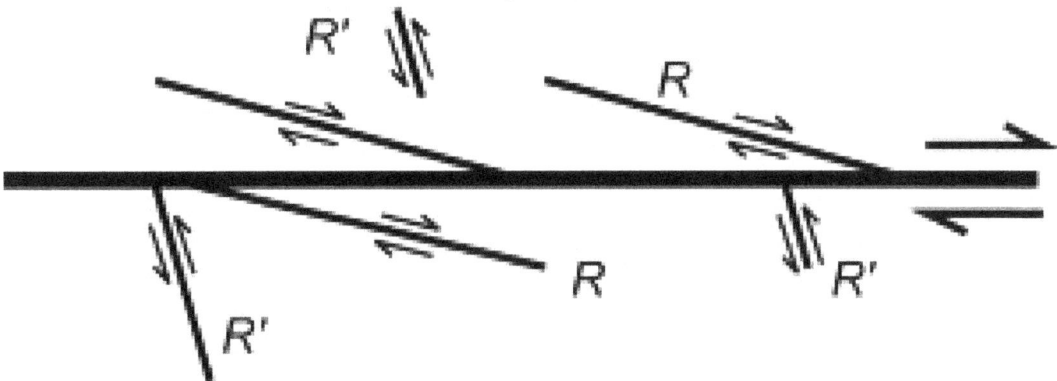

Figure 10. Riedel shears associated with a dextral fault.

Fault-bend folds. If a fault is curved, movement of the fault inevitably causes bending of either the hanging wall or the footwall or both. (If this did not happen, caverns would open up within the Earth as faults moved; lithostatic pressure prevents this from happening.) Fault-bend folds are particularly common in thrust belts like the Canadian Rocky Mountains.

Figure 11. Fault-bend folds above thrust fault in siderite bed, Stellarton, Nova Scotia.

Detachment folds. Sometimes, the amount of slip on a fault varies across the fault surface. Some sections have moved more than others. At the boundary between the sections, deformation in the wall rocks becomes intense, sometimes leading to the formation of **detachment folds.**

Figure 12. Detachment fold, Mount Rundle.

Drag folds. Sometimes, a fault will 'lock' during its development, and if this happens deformation may be spread through the wall rocks. If conditions are right, this deformation may be ductile and features in the wall rocks may be bent so that they become folded. Such folds are called **drag folds**. Drag folds can closely mimic fault propagation folds (below) that are formed at the tip of a fault as it develops. Many folds formerly categorized as drag folds are now interpreted as fault-propagation folds.

Figure 13. Drag fold.

Faultpropagation folds emerge during fault development, incorporating characteristics of both fault-bend and detachment folds.As a fault progresses, a fault tip will emerge at the juncture between displaced rock and rock that remains stationary.Faultpropagation folds develop at the fault tip, resulting in the shortening of one fault wall in relation to the other. They are also prevalent in thrust belts such as the Rockies.

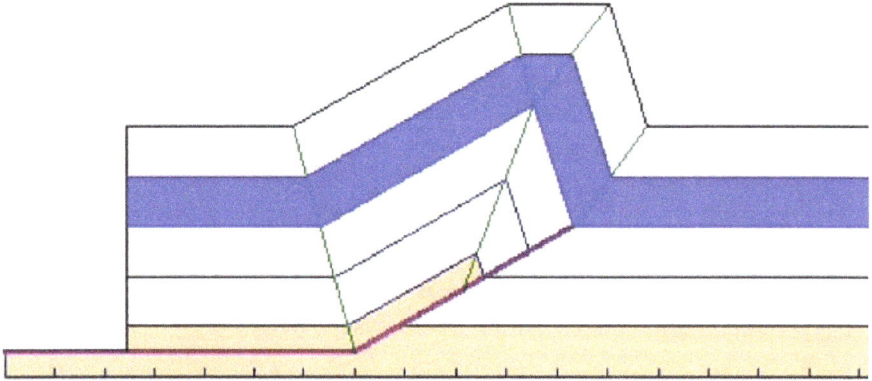

Figure 14. Fault propagation fold (based on a computer simulation by H. Charlesworth).

Figure 15. Fault propagation fold, Arisaig, Nova Scotia.

LAB 8. MEASURING FAULT SLIP

In Lab 7 you dealt with fault constructions that involved the separation of surfaces, but not with fault slip. This is because cut-offs for a single plane surface are inadequate to determine the **displacement vector**, or **slip** of a fault. In this lab you will analyse areas where **piercing points** can be located in both the footwall and hanging wall of a fault, allowing the true slip to be determined.

Assignment

1. * Map 1 shows a simple map of a level landscape 500 m above sea level, in which a fault offsets a mafic dyke with a strike separation of 450 m. Slickenlines on the fault trend toward the south and have rake of 060°. Determine the slip of the fault.

   a) *Plot the fault and the dyke as great circles on a stereographic projection. Also plot the slickenline lineation, and the intersection line, where the fault and dyke intersect. Use a small 'x' for the linear data.

   b) *Now draw a fault-plane cross-section, viewed from the southwest. This section will show what you would see if you were able to remove the hanging wall block completely, and look directly at the fault face on the footwall. The land surface should be a horizontal line at the top of the section. The cutoff-line of the dyke in the footwall should be shown with its correct rake. Draw a few of the slickenlines, also with their correct rake.

   c) *Now mark, using a dashed line, the cutoff of the dyke in the hanging wall (450 m across strike from its position in the footwall). Draw a line between the two cutoffs, parallel to the slickenline orientation. This is the net slip. Measure the distance of net slip and determine its trend and plunge from the stereographic projection. Say whether the dip-slip component is normal or reverse. Say whether the strike-slip component is sinistral or dextral.

2. Map 2 shows an area in which a thick sequence of conglomerate (circles) unconformably overlies mudstone (white) and coal (black). The area is cut by a major fault running NE-SW approximately. To the east of the fault the coal is not exposed, but three boreholes at A, B, and D encountered the coal at depth. Borehole C did not encounter coal. Your objective is to determine the subsurface structure, so as to show the extent of coal in the subsurface. In addition, you are to define the slip of the fault, using the subcrop of the coal below the unconformity.

Borehole	Elevation of Ground	Depth to Unconformity	Depth to Coal
A	480	180	280
B	513	213	433
C	650	350	absent
D	580	280	310

a) Fault plane

i) Draw structure contours on the fault surface itself. Extrapolate contours down to sea level (0 m).

ii) Determine the strike and dip of the fault.

b) West of the fault the coal is exposed

i) Identify the unconformity by labelling it on the map.

ii) What is its orientation?

iii) Draw structure contours on the coal seam (assume that the coal seam is thin relative to the scale of the map, and therefore treat it as a single surface). Terminate the contours where they intersect the fault.

iv) What is the orientation of the coal?

v) Mark the **subcrop** line on the map where the unconformity surface cuts the coal seam.

c) East of the fault the coal is not visible

i) Identify the unconformity by labelling it on the map.

ii) What is the orientation of the unconformity?

iii) The borehole data allow the elevation of the coal seam to be calculated at three points (A, B & D). Use this information to construct structure contours on the coal seam.

iv) From your contours, determine the strike and dip of the coal east of the fault.

v) Mark the **subcrop** line where the unconformity surface cuts the coal seam.

d) Separation: Next, investigate the separation of the surfaces that are cut by the fault.

i) Using intersecting contours, mark and label the **cutoff lines** where the fault intersects the coal seam in the footwall and in the hanging wall. Between these two lines is a region where a drillhole would encounter no coal.

ii) Find the difference in elevation between the unconformity in the west and the same unconformity in the east. This is the **vertical separation** of the unconformity at the fault – the vertical distance between a surface and its projected counterpart from the other side of the fault. Because the surface is horizontal, it is also the **throw** (vertical distance between the hanging wall and footwall cutoffs, measured down the dip of the fault plane).

iii) Find the difference in elevation for the coal seam on either side of the fault (the **vertical separation)**. (Note: Because the coal seam is gently dipping, its separation measurements – vertical separation, throw, and heave are all slightly different from the corresponding measurements made on the unconformity, which is horizontal.)

iv) Find the distance measured along the strike of the fault, between equivalent structure contours on the coal seam on either side of the fault. For example you might find the point where the east side 300 contour hits the fault plane, and the point where the 300 contour on the west side hits the fault plane. This is the **strike separation** of the fault. It corresponds to the distance that would appear between the two halves of the coal seam if the land were eroded down to a horizontal surface.

v) Measure the **heave** of the fault at the coal seam, the width of the zone (measured perpendicular to the strike of the fault) between the two cutoff lines.

vi) Superimpose a sheet of tracing paper on the map; on it, shade the area where the coal seam is present below ground level. (*Hint:* To do this, the best argument is as follows. Originally the coal was present under the whole area. Since deposition, coal has been 'removed' in three ways. First, some coal was removed by erosion prior to deposition of the conglomerate, at the unconformity surface; second, coal is missing from the fault heave between the two cutoff lines; third, more coal has been removed by erosion to the present-day topography. Once you have removed all these areas, whatever is left is the area underlain by coal.)

e) Fault slip: So far, all the measurements you have made concern the separation of various surfaces. Although they would be very useful to an exploration project seeking the location of the coal seam, they do not tell us the actual movement on the fault, the net slip. To determine the net slip, it is necessary to find a unique line which has been offset. There is an identifiable line that has been severed by the fault. This is the line where the unconformity intersects the coal seam, the **subcrop** of the coal seam. This line has been split into two parts by the fault, creating **piercing points** on both walls. It can be used to pin down the true slip.

i) Set up a fault-plane cross-section.

◉ Use a piece of graph paper to make a fault plane cross-section, using the sea-level contour as a folding line. The trace of the fault on the map is somewhat longer than a normal letter-size sheet of tracing paper, but you will be able to show all the important intersections on a cross-section 25 cm (6.25 km) long, starting at the southern boundary of the map. First, find the point where the sea-level (0 m) contour on the fault plane intersects the south border of the map. Call this point X. Measure 25 cm NE along the contour and mark point Y. Now take a sheet of graph paper and draw a horizontal line 25 cm long, about 5 cm up from the base of the paper, and mark the ends X and Y.

◉ Next calculate the spacing of contours on the fault plane. If the fault has dip d, the contour spacing on the section will be: $i' = i / \sin(d)$

ii) Add the footwall and hanging wall intersections.

◉ Place your graph paper with its top edge along the line XY on the map, and mark the places where the footwall and hanging wall cutoff lines have known elevation. Project these points to their correct elevations on the fault-plane section.

◉ Complete the cross-section by drawing the hanging wall and footwall traces of the unconformity and the coal. Use solid lines to represent the footwall cutoff and dashed lines to represent the hanging wall cutoffs.

iii) Determine the fault slip:

◉ Identify the piercing points where the subcrop line intersects the fault, in both the hanging wall and footwall. Join the two points with a dash-dot line representing the **net slip**.

◉ Measure and record the distance of net slip.

◉ The component of net slip parallel to the fault strike is the **strike slip**. Draw and label the strike slip on the fault-plane section, and record the distance of strike slip.

◉ The component down the dip of the fault is the **dip slip**. Draw the dip slip, and label it. Record the amount of dip slip.

◉ Characterize the fault in words overall, as mainly strike-slip, intermediate (oblique-slip), or dip-slip. Say whether the strike-slip component is sinistral or dextral. Say whether the dip-slip component is normal or reverse.

f) Check the results on a stereographic projection

i) Find the rake of the net slip in the fault plane, by measuring it with a protractor on the fault-plane section.

ii) Plot the fault plane, the orientation of the coal, the cutoff lines, and the net slip on a stereographic projection, and determine the plunge and trend of the net slip.

Answer table for fault measurements, Question 2

Strike and dip of the fault:	
Orientation of the unconformity west of the fault:	
Orientation of the coal seam west of the fault:	
Orientation of the unconformity east of the fault:	
Orientation of the coal seam east of the fault:	
Vertical separation of the unconformity at the fault:	
Vertical separation of the coal seam at the fault:	
Strike separation of the coal seam at the fault:	
Heave of the fault at the coal seam:	
Net slip:	
Characterize the fault slip:	

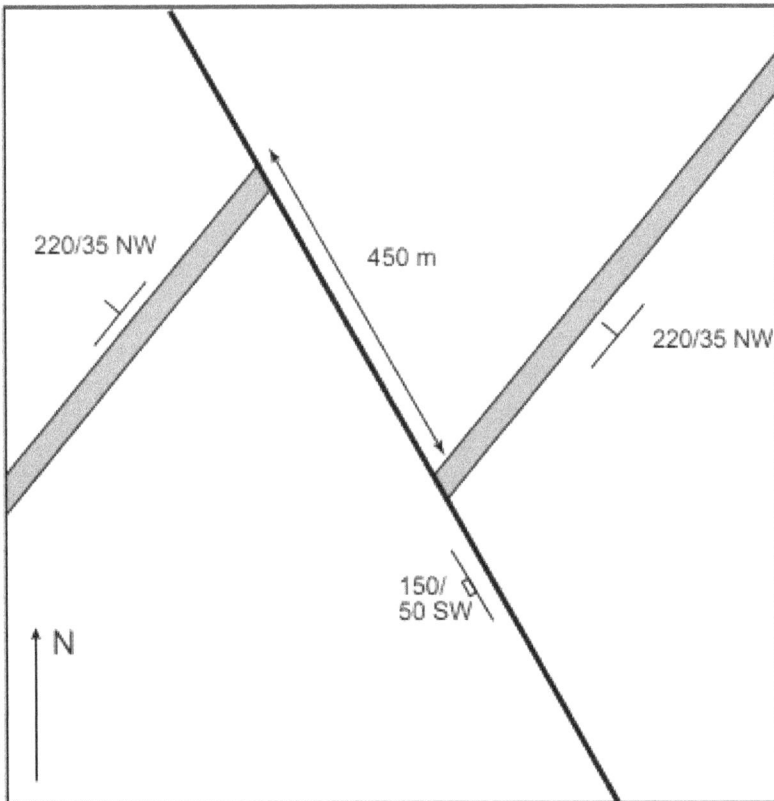

Lab 8. Map 1. [PDF]

Lab 8. Map 2. [PDF]

12 Chapter | TECTONIC ENVIRONMENTS OF FAULTING

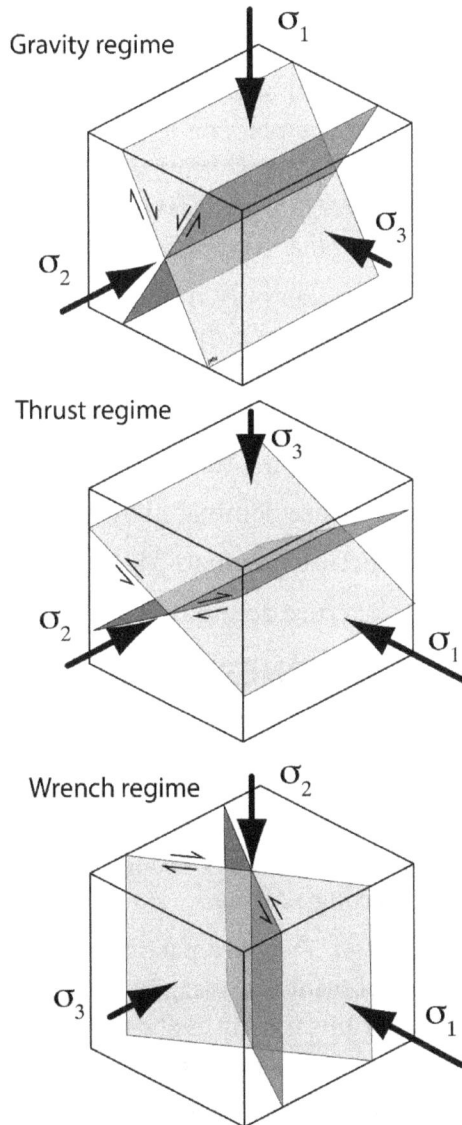

Figure 1. Typical relationships between stress and faults based on Anderson's (1960) analysis of near-surface states of stress.

DYNAMICS OF FAULTING

In contrast to ductile structures, fractures can be experimentally recreated using actual rock samples subjected to stress in presses. Figure 1 illustrates the typical correlation between shear fractures and stress. These fractures are referred to as conjugate shear fractures. Conjugate shear fractures present one of the rare instances in which structural geologists can confidently derive dynamic interpretations from straightforward field evidence. Anderson's (1905) theory of faulting starts from some basic facts about stress, and leads to a classification of tectonic environments into **fault regimes.** Anderson noted that the Earth's surface is effectively a plane of zero shear stress – at least as far as structural geologists are concerned. (We can neglect the relatively tiny shear stresses due to the wind and the flow of water.) This means that the Earth's surface closely approximates a **principal plane of stress**, and the pole, or normal, to the Earth's surface must be one of the **principal stresses**. Because the Earth's surface is regionally close to horizontal, this means that over large areas of the Earth's surface, *one of the principal stresses is vertical, and the other two are horizontal.*

Note that this argument strictly applies only at the Earth's surface. Once we descend to lower levels in the lithosphere, the principal stresses can, and do, wander away from their vertical and horizontal orientations.

Based on which principal stress is near-vertical we can classify environments in the Earth's upper crust into the following three categories.

Gravity régime: σ_1 vertical: structure dominated by normal faults.

Thrust régime: σ_3 vertical: structure dominated by thrust faults.

Wrench régime: σ_2 vertical: structure dominated by strike-slip faults.

RIFTS AND EXTENSIONAL ZONES: NORMAL FAULTS

Occurrence

Several parts of the Earth's crust are undergoing horizontal extension and crustal thinning at the present day. The best studied examples are probably the Basin and Range province of the western USA, and the East African Rift valley, which extends several thousand kilometres through Africa.

We can find ancient examples too. They are particularly well studied underneath **passive continental margins**, continental margins which are currently not located at a plate boundary. Passive margins mostly evolved from rift valleys similar to the East African Rift, as continents stretched and separated in the past. For this reason we find rift structures buried beneath passive continental margins in many parts of the world. Many of these contain significant petroleum reservoirs. The **Hibernia structure** on Canada's east coast, formed in a rift that developed as the Atlantic opened. Many of the oilfields in the European **North Sea** occupy structures related to the **Viking-Central Graben,** buried beneath younger sedimentary rocks in the North Sea.

Features of normal faults in rifts

These regions are dominated by normal faults, and in many rifts they have a roughly conjugate geometry: faults dipping towards and away from each other with dips around 60°, at least near the surface. In many cases faults can be shown to flatten at depth, giving them **listric** geometry, though in some cases this is controversial. Some cross-sections of rifts show faults dipping at a constant angle down to a subhorizontal shear zone in the brittle crust at depth.

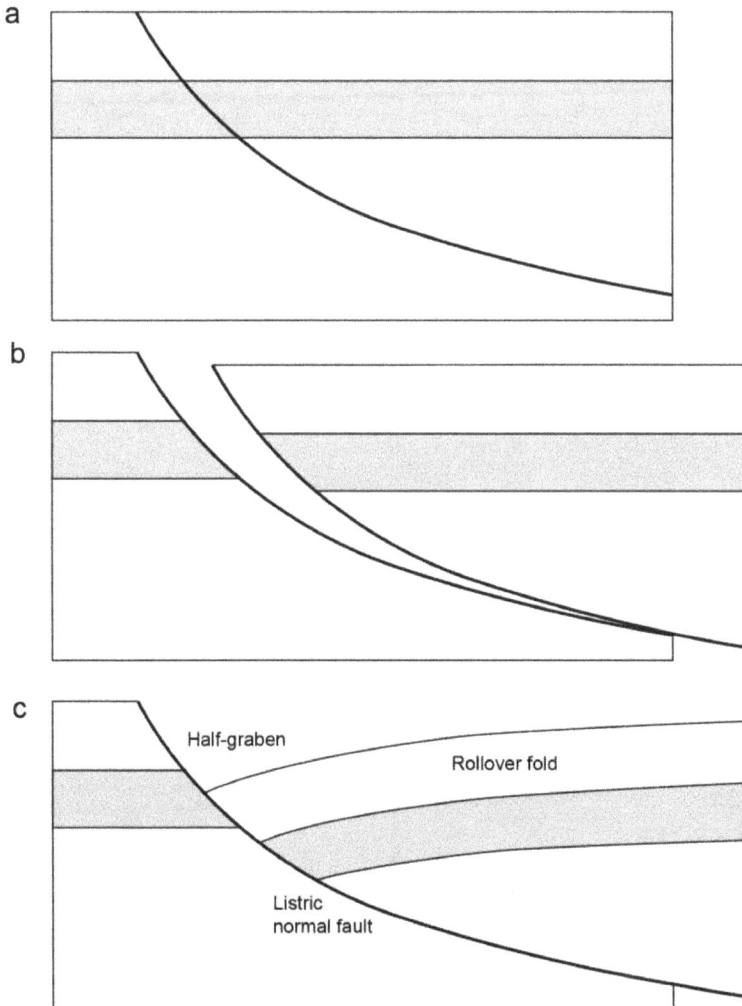

Figure 2. (a) Listric normal fault. (b) Displacement of hanging wall without folding would lead to a large open space. (c) Hanging wall deforms by fault-bed folding to produce a rollover anticline.

Listric shape has consequences for the geometry of the walls of faults: one or both walls must distort to maintain *strain compatibility*. Typically, the hanging wall distorts more (because it has a free surface at the top), and displays a characteristic type of fault-bend fold called a **rollover anticline**. Rollover anticlines form important petroleum traps in many passive continental margins.

Other factors may lead to folding of the wall rocks. It is common in sedimentary basins for slip to vary across the surface of a fault, so that slip declines at the edges of a mapped fault, to a point where there is no slip at all, called the **tip** of the fault. How is this accommodated in the wall rocks? Typically, slip is transferred to another fault, but if the two faults are not connected, distortion of the wall rocks may produce a **relay ramp** linking the two.

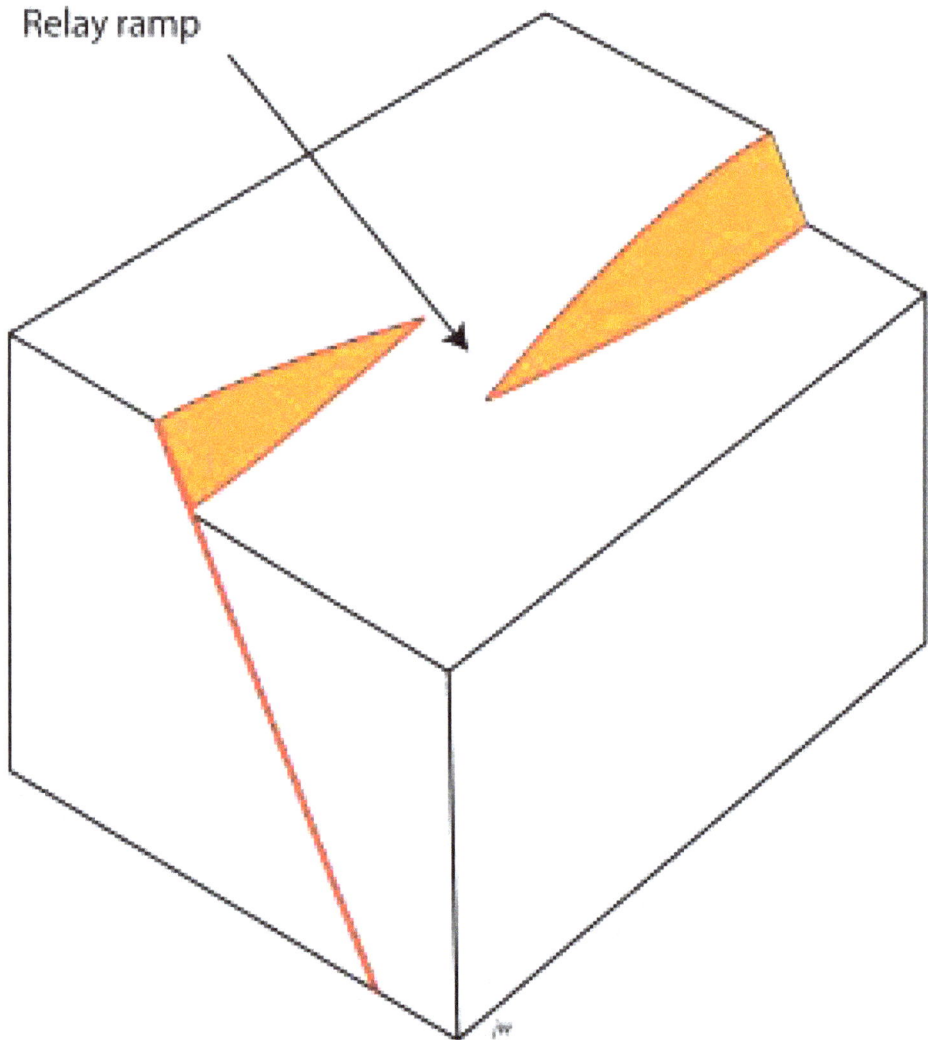

Figure 3. Relay ramp between the tips of two normal faults.

Arrays of normal faults

Normal faults in rifts do not occur in isolation. Often a cross-section through a rift will show dozens or hundreds of normal faults. Special terms are given to features of normal fault arrays.

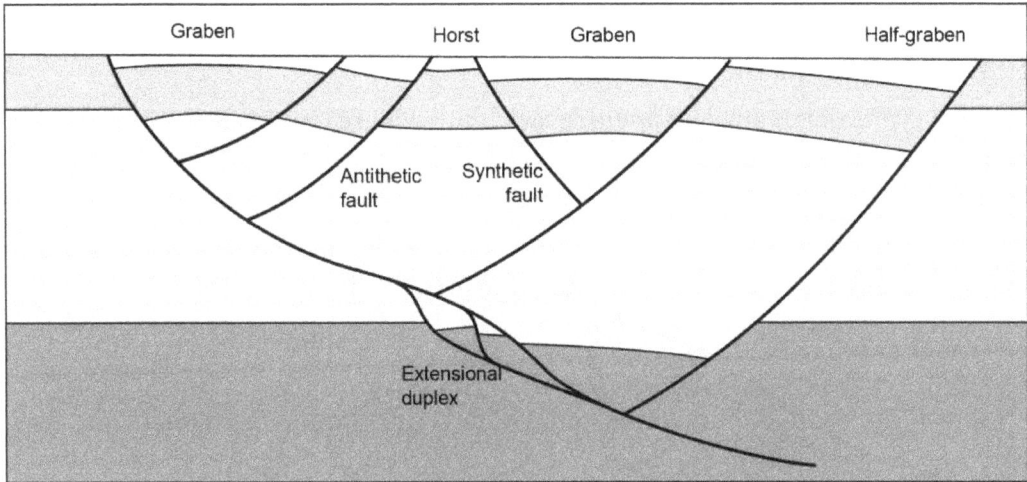

Figure 4. An array of normal faults in a rift.

A **horst** is a block between two normal faults that dip away from each other, in opposite directions. The rocks in a horst stand higher than the rocks on either side.

A **graben** is a block between two normal faults that dip towards each other. The rocks in a graben stand lower than the rocks on either side.

A **half-graben** is a tilted block in a rift zone, that is also dropped down relative to the rocks on either side, but it only has a fault on one side; the other side is typically gently folded.

In many rifts there are a few dominant normal faults with very large offsets. Faults that dip in the same direction are described as **synthetic** to the main faults; faults that dip in the opposite direction are **antithetic**.

THRUST BELTS

Occurrence

Areas of crustal shortening, dominated by reverse faults, are active at the present day at convergent plate boundaries (*subduction zones*), and in areas of *continental collision*: such as the Himalayas, and Taiwan.

Ancient examples occur in the Rocky Mountains formed in the late Mesozoic and early Cenozoic eras, and the Appalachian Mountains of eastern North America, formed in the Paleozoic era.

In most of these examples folds are intimately associated with thrusts.

In many cases, there is a sedimentary basin, called a **foreland basin**, adjacent to the thrust belt, and it is thought that the weight of the thrust belt has depressed the lithosphere as it formed by *isostasy*. In western Canada, most of the Cretaceous rocks formed in such a basin, including those that we see in the river valley in Edmonton.

Features of single thrust faults

Geometry: A thrust is any low-angle reverse fault, where low-angle is rather loosely defined. In fact, some thrust faults are sufficiently curved so that portions would technically be normal faults!

Because of their low angle character, thrusts may look like stratigraphic contacts on maps. There may be closed outlines of thrusts at hills and valleys.

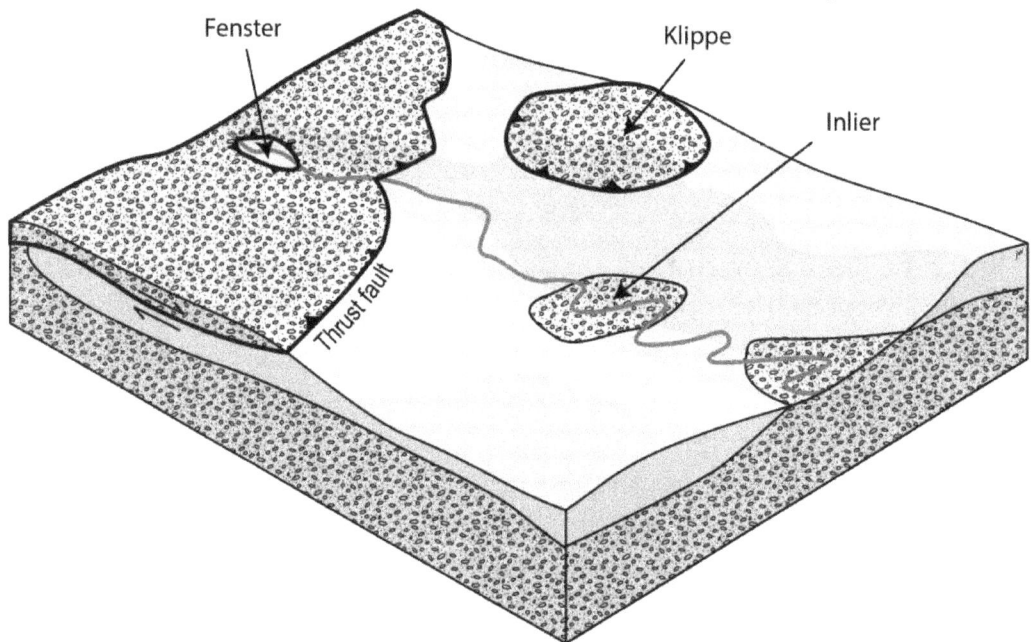

Figure 5. Features of thrust faults in thrust belts.

◉ An area of hanging wall entirely surrounded by footwall is called a **klippe.**

◉ An area of footwall entirely surrounded by hanging wall is called a **fenster** or **tectonic window.**

Thrusts are influenced by stratigraphy. In many instances, the dip is steeper in *competent* rocks like limestone and sandstone, and shallower in *incompetent* rocks like shale and evaporites (rock salt, gypsum, anhydrite).

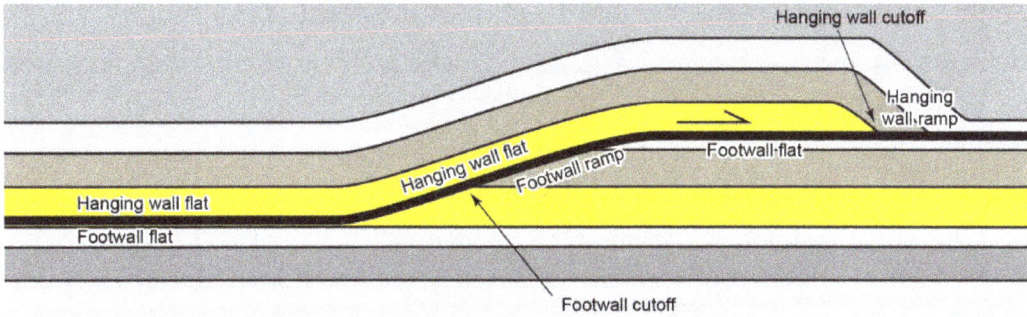

Figure 6. Single thrust fault, showing fault-bend folds developed over ramps and flats.

The steeper portions are **ramps.** They can be recognized on maps and cross-sections because the fault cuts stratigraphic boundaries at a relatively high angle. On ramps we can find cutoff points. Understanding the kinematics of a thrust belt often involves matching footwall and hanging wall cutoffs on the same fault. This process is an important part of checking cross-sections through thrust belts, known as **section balancing.**

The gentler portions are **flats.** They can be recognized on maps and cross-sections because the fault surfaces are almost parallel to the stratigraphic boundaries.

Notice that once a fault has moved, there will be *separation* between the **hanging wall ramps** and the **footwall ramps,** and between the **hanging wall flats** and the **footwall flats.** If fault-bend folding occurs, then the hanging wall flats may acquire steeper dip and the hanging wall ramps may flatten, so this terminology may seem strange! They can still be recognized by the characteristic *cutoff angles* of the strata.

A very extensive flat, where there has been a large amount of movement, is called a **décollement** surface, from a French word meaning 'unstick'. If the hanging wall has moved a long way relative to the footwall (tens of kilometres or more), far enough that the hanging wall rocks were formed in a completely different environment from the footwall, then the hanging wall is described as an **allochthon** (from two Greek words, meaning "other place") and is said to be **allochthonous**. The footwall of such a thrust is called the **autochthon**, and is said to be **authochtonous.**

For example, in Newfoundland there is a complex of thrust sheets known as the **Humber Arm Allochthon** that is composed of Cambrian and Ordovician deep-water sedimentary rocks and ophiolites from the floor of the Iapetus Ocean. It was thrust on top of sedimentary rocks of more or less the same age range that were formed on a continental shelf, and therefore represent a very different environment.

Folds associated with thrust faults

Folds may form in association with thrust faults. A wide variety of fold styles are recognized but we will recognize just three types:

Fault-bend folds: formed where the dip of the fault changes;

Detachment folds: formed where the slip changes, and distortion is accommodated in the hanging wall;

Fault-propagation folds: formed where the dip and the slip change, typically where a fault tip climbs up a ramp during fault-propagation.

In reality there are many variations on these themes. In particular, there is probably a continuous variation between the idealized detachment and fault propagation folds.

Arrays of thrust faults

Vergence is a useful concept in dealing with arrays of thrusts. The vergence of a structure is the direction in which rocks near the surface have moved relative to rocks deeper down. In some thrust arrays, all the thrusts have the same vergence: that is to say the hanging walls have all been displaced the same way relative to the footwall. Large parts of the Rocky Mountain thrust belt show eastward vergence, for example. The **foreland** is the area towards which most of the thrusts verge: in the Rockies, that is the plains of Alberta. The opposite side of the thrust belt, the side from which most of the thrust sheets appear to come, is called the **hinterland.**

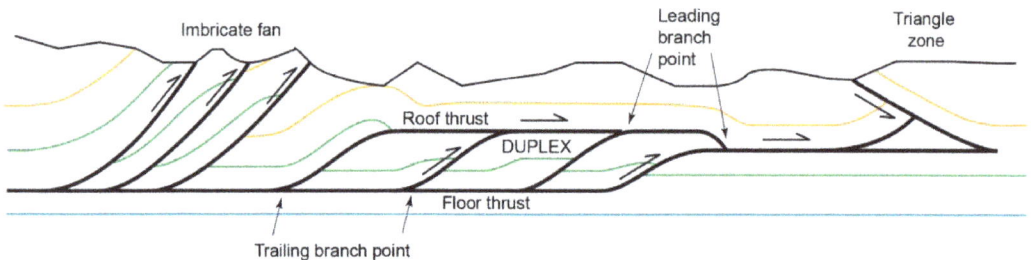

Figure 7. An array of thrust faults in a thrust belt.

Thrust arrays with similar vergence

Once we start looking at multiple thrusts things can get quite complicated. It is common for multiple listric thrust faults to branch upward from a single flat. This configuration is called an **imbricate fan.** The point where one thrust fault branches from another is called **a branch point.** In an imbricate fan the branch points mark places where one fault branches into two as you follow it towards the foreland. This type of branch point is a **trailing branch point.**

Sometimes imbricate thrust faults merge into flats both upward and downward. Such a configuration is called a **duplex.** The upper flat is a **roof thrust** while the lower flat is the **floor thrust** of the duplex. Like an imbricate fan, the thrusts in a duplex join downwards at trailing branch points. However, they also join upwards at **leading branch points.**

Thrusts with opposing vergence: wedges and triangle zones

Sometimes faults in thrust belts have opposing vergence. A pair of opposed thrusts that meet in the subsurface is called a **tectonic wedge**. If a third thrust cuts off a triangular section of rock, the wedge is called a **triangle zone**. The earliest oil discoveries in the foothills of the Canadian Rockies were found in a triangle zone in the Turner Valley area of southern Alberta.

STRIKE-SLIP ZONES

Occurrence

Some of the Earth's best-known faults are strike-slip faults: the San Andreas in California is one of the best-known and best studied. Other strike-slip faults that have histories of damaging earthquakes include the North Anatolian Fault in Turkey and the Alpine Fault in New Zealand. Strike-slip faults that are also plate boundaries are called **transform faults**.

Characteristics of strike-slip faults and fault zones

Anderson's theory of faulting predicts that strike-slip faults should be characteristic of the **wrench regime**, and that those faults should be near-vertical. Most major strike-slip faults are found to dip steeply, confirming Anderson's prediction. As a result, their traces tend to be very straight lines on geological maps.

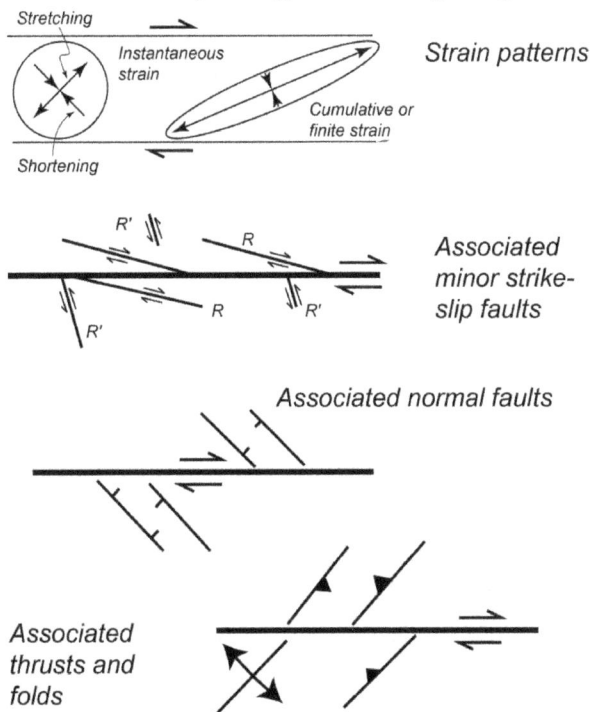

Figure 8. Structures associated with strike-slip faults.

The damage zones around strike-slip faults tend to contain a mixture of structures that we normally associate either with horizontal shortening (as in thrust belts) or with horizontal extension (as in rifts). Cross-sections through strike slip faults tend to be quite confusing, for this reason. The best way to show what's going on in a strike-slip zone is in map view.

In map view, a strike-slip fault zone may be idealized as a zone of simple shear. For the sake of argument we will draw a dextral strike-slip zone. (For a sinistral strike-slip zone everything would be in mirror image.) In a dextral strike-slip a zone, it's possible to show that the most rapid shortening takes place along a line 45° clockwise from the shear-zone boundary, and the most rapid extension takes place along a line 45° counter-clockwise from the shear-zone boundary. For a sinistral strike-slip zone, the opposite senses of rotation apply.

This idealized pattern of strain rates leads to some predictions of how structures will form. We predict that extensional structures like joints, veins, and normal faults, will form perpendicular to the extension direction, and shortening structures like folds, thrust faults, and cleavage planes will develop perpendicular to the shortening direction.

It is also quite common to see subsidiary strike-slip faults, which may include Riedel shears. Dextral **synthetic Riedel shears** would be predicted at about 15° clockwise from our overall dextral strike-slip zone, while sinistral **antithetic Riedel shears** would be about 75° clockwise from the overall direction of dextral strike slip. (Reverse everything for a sinistral zone.)

All these orientations work for fault zones where the overall amount of strain is small. Strike-slip deformation includes a strong component of rotation, so as deformation continues, all the structures will be rotated. One of the challenges in interpreting strike-slip motion is that structures may have rotated out of the orientations in which they formed.

Transtension

The above predictions apply to ideal strike-slip motion. However, many of the most interesting structures in strike-slip zones are formed where there are departures from ideal strike-slip.

For example, a combination of strike-slip motion with extension is called **transtension**. In transtension there is a component of crustal thinning, along with strike slip, so transtension zones tend to subside and form sedimentary basins.

One common environment of transtension is at a **releasing bend** (also known as a **releasing stepover)** on a strike-slip fault. If a dextral fault steps to the right, or a sinistral fault steps to the left, the resulting bend is said to be **releasing**, and the rocks adjacent to the fault are affected by transtension.

A releasing bend typically develops a localized, parallelogram-shaped subsiding area called a **pull-apart basin.** In the Los Angeles area, pull-apart basins associated with the San Andreas system host important natural resources of oil and natural gas. In Nova Scotia, the Stellarton Basin, also a pull-apart basin on a dextral strike-slip fault system, was a prolific producer of coal in the 20th century.

In cross-section, pull-apart basins tend to be bounded by families of faults that steepen downwards and merge into a single fault or shear zone at depth. Individual faults may have normal, strike slip, or oblique slip. This type of fault array is called a **negative flower structure.**

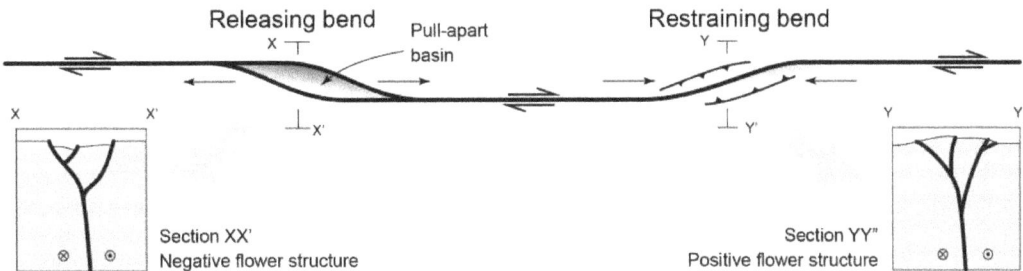

Figure 9. Transtension and transpression at releasing and restraining bends on a dextral strike-slip fault.

Transpression

A combination of strike-slip motion with shortening is called **transpression**. In transpression there is a component of crustal thickening, along with strike slip, so transpression zones tend to form narrow uplifts, ranges of hills or mountains.

One common environment of transpression is at a **restraining bend** (also known as a restraining stepover) on a strike-slip fault. If a dextral fault steps to the left, or a sinistral fault steps to the right, the resulting bend is said to be restraining, and the rocks adjacent to the fault are affected by **transpression**.

A restraining bend typically develops a localized uplift. Parts of the transverse ranges of California, and the Southern Alps of New Zealand, are associated with transpression at restraining bends along major transform faults.

In cross-section, transpressional ranges tend to be bounded by families of faults that steepen downwards and merge into a single fault or shear zone at depth. Individual faults may have reverse, strike slip, or oblique slip. This type of fault array is called a **positive flower structure.**

Anderson, E. W. 1905. The dynamics of faulting. *Transactions of the Edinburgh Geological Society* **8**, 387-402.

LAB 9. FIELD MAPPING

Introduction: An Overview

Field geologists frequently require the ability to document the relationships among a series of outcrops to gather the most pertinent information for geological map creation in the least amount of time. Geologists engaged in mapping consistently possess a base map for documenting their observations. For mapping large areas this might be a topographic map at any scale between 1:10,000 and 1:50,000. For a small area it may be a sheet of gridded paper on which the geologist records both geologic and topographic information. Recording information directly on the map while in the field is essential; that way, cross-cutting relationships such as faults, intrusions, and unconformities can be swiftly and directly portrayed on the map. Also, you will find out whether you need to collect more evidence for a critical relationship before you leave the field. Modern methods of surveying, particularly the use of the Global Positioning System (GPS), have greatly assisted map-making at large scale. However, portable GPS units often give errors of 5 m or more in location. Detailed field relationships may still need to be surveyed using tape-and-compass or pace-and-compass methods.

This exercise will take place at a location where you can practise mapping techniques with a variety of rock types and structures. Your instructors or teaching assistants will show you the area to be covered. For this exercise you will need a notebook, a clipboard and a sheet of graph paper, coloured and lead pencils, a compass-clinometer, and your legs! You will also need to be appropriately dressed and equipped for working outdoors. For all geological fieldwork, it is important to carry clothing and equipment appropriate to the range of possible conditions you may encounter.

At the University of Alberta, this lab will take place in the Geoscience Garden, a facility that is set up to enable you to practise mapping techniques with a variety of rock types and structures. Despite its nearby location, you will still need to be prepared for work outdoors. The weather in Edmonton can be unpredictable. You will probably need gloves and you may also need a waterproof coat and footwear. Alternatively, if it is sunny you may need sunscreen and a hat.

Make sure that your compass-clinometer is correctly set for magnetic declination at your location.

The Geological Survey of Canada has a useful declination calculator at:

Magnetic declination calculator: https://geomag.nrcan.gc.ca/calc/mdcal-en.php

For example, in Edmonton AB, Canada, at the beginning of 2020, the declination was 13.8° East

Assignment

The mapping area is large and contains a wide variety of rocks. Your teaching assistants will designate an area for the mapping exercise.

Measure your pace

1. * When mapping, it is helpful to be able to estimate short distances by pacing. A tape measure will set up on to extend 50 m or more through the mapping area. Pace along the tape at least three times to establish the approximate length of your pace. The most convenient method is to count one for each double pace (one-and-two-and-three...). In other words count a pace each time your left foot hits the ground. Pace naturally as you would while walking. Don't stretch or try to pace extra long steps.

    Divide 50 by the average number of paces to get the length of your pace. Round it to a convenient number like $1\frac{1}{2}, 1\frac{1}{3}, 1\frac{2}{3}$, etc., so that you can approximately convert paces to metres in your head (measurements approximated to the nearest 1 m are quite good enough).

    Also, measure the trend of the tape, and if your basemap has topographic features, locate the tape by pacing from one end of it to a topographic feature.

2. * Using the scale of your map, draw a line on the graph paper to represent the tape.

Detailed description of first outcrop

3. Choose one outcrop of sedimentary rock and make a detailed description of the rock type(s) as an entry in your notebook. Give the entry a number. Your description should include grain size, grain shape (if visible), composition (if you can tell), primary structures (bedding, lamination, cross-lamination, etc.), and secondary structures (folds, fractures, fabrics). Points to remember in making good field notes:

⊙ Print your notes so that anyone can read them;

⊙ An outcrop may contain more than one type of rock. Make sure that your notes capture the variation from one lithology to another;

⊙ The degree of detail will depend on the rock-type. For example, in a conglomerate it's possible to make a very full description of the clasts, whereas in a mudstone you probably won't be able to directly observe the grain shape, sorting, and composition. Don't claim to observe features you can't see!

4. Measure the orientation of any layering you see, and identify the type of layering (lamination, bedding, gneissic banding, etc.

5. Make a diagram of a any evidence that tells you whether the beds are upright or overturned. Points to remember in making a good field diagram:

⊙ Draw what you actually observe: it's better to draw a small part of the rock well rather than a large part showing what you think you should see;

⊙ Your diagram does not have to be artistic. In the field you can take a photo to supplement your diagram, but..

⊙ Your diagram must have labels to indicate what is represented. (You can't include these on a photo, so even if you have a camera, making a labelled diagram is still important.)

⊙ Make sure you include an approximate scale bar;

⊙ Include an indication of the orientation of the view, that shows whether you are looking at a plan or a cross-section.

6. Locate the outcrop on the map; if necessary, draw its outline at approximately the right size and shape. Lightly shade the outcrop with a colour. Mark a strike-and-dip symbol in approximately the right orientation beside the outcrop. Try to place the symbol along strike from the outcrop it describes.

Map observations

7. Locate and describe the other outcrops in the designated area of the garden. Number your notebook entries and mark the locations on the map. When the rock types are similar, your descriptions can be brief; you need only describe major differences. However, you should try to measure at least one strike and dip at each outcrop. In the course of mapping you may encounter folds or faults; measure the orientations of these structures.

8. You will encounter a variety of rock types. Make sure you make a full description of each new rock type in your notebook. Try to group the rocks you observe into *formations.* Give each new formation a distinct colour, and build a legend to the colours on a spare part of the map or in your notes.

Interpreting boundaries

9. As you encounter differences between rock types, try to mark where boundaries might be using dashed lines. Remember, boundaries between sedimentary rock units tend to be parallel to the strike. After you have decided where boundaries are, very lightly shade the area where you think each formation is present between the observable outcrops.

Do not leave the field area until your map shows an interpretation of the geology throughout the are, with clear interpreted boundaries between units. You can judge the effect of topography much better in the field than in the lab, so complete the map in the field.

Your map must be approved by an instructor or teaching assistant before you leave the field.

After leaving the field

10. On your return to the lab, make a good copy on tracing paper. Mark symbols for the strike and dip of bedding, and any other structures, this time by measuring them accurately. If you cannot place the symbol exactly on the outcrop without making the map confusing, offset it slightly to the side, shifting it in the direction of its own strike or trend.

11. If you encountered a fault or faults, use your observations at a fault to determine the approximate **slip**, using the same method you used in Lab 8. If you think the area contains a fold, determine the orientation of the **fold axis** and **axial surface**, using the same method you used in Lab 6.

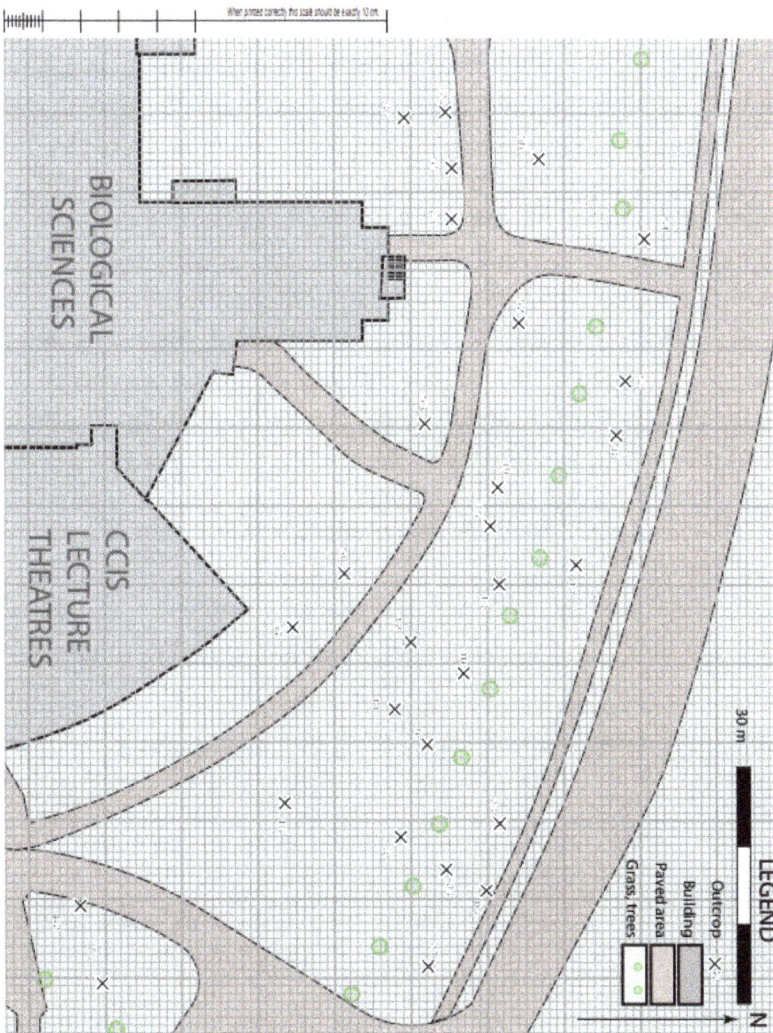

Lab 9. Base map [PDF]

LAB 10. FOLD AND THRUST BELTS

Introduction: An Overview

The Foothills, Front Ranges, and Main Ranges of the Canadian Cordillera constitute one of the most thoroughly characterized fold and thrust belts globally, where sedimentary strata have been compressed by hundreds of kilometers while maintaining moderate temperatures and exhibiting, at most, modestgrade metamorphism.Additional instances of fold and thrust belts encompass the Appalachian Valley and Ridge province in the eastern United States, the Jura Mountains in the Swiss Alps, and the Moine Thrust zone in northwest Scotland. Numerous belts possess significant stores of oil and natural gas within structural traps, prompting extensive efforts to comprehend their geometry and kinematics. The comparatively low temperatures during rock deformation result in significant disparities between competent lithologies, such as limestones and sandstones, and incompetent lithologies, such as shales and evaporites.Consequently, faults typically exhibit rampflat courses within the strata, and the majority of folds are associated with variations in the slip or dip of faults, or both.Moreover, competent lithologies are often transported over considerable distances with minimal internal distortion, particularly at the hinges of folds.Consequently, fold and thrust belts frequently exhibit layers that preserve their original thickness; such folds are referred to as parallel folds (Fig. 1).These qualities are advantageous in constructing crosssections, particularly in regions where the abundance of faults and folds complicates traditional structure contouring.

IDEALIZED FOLD STYLES

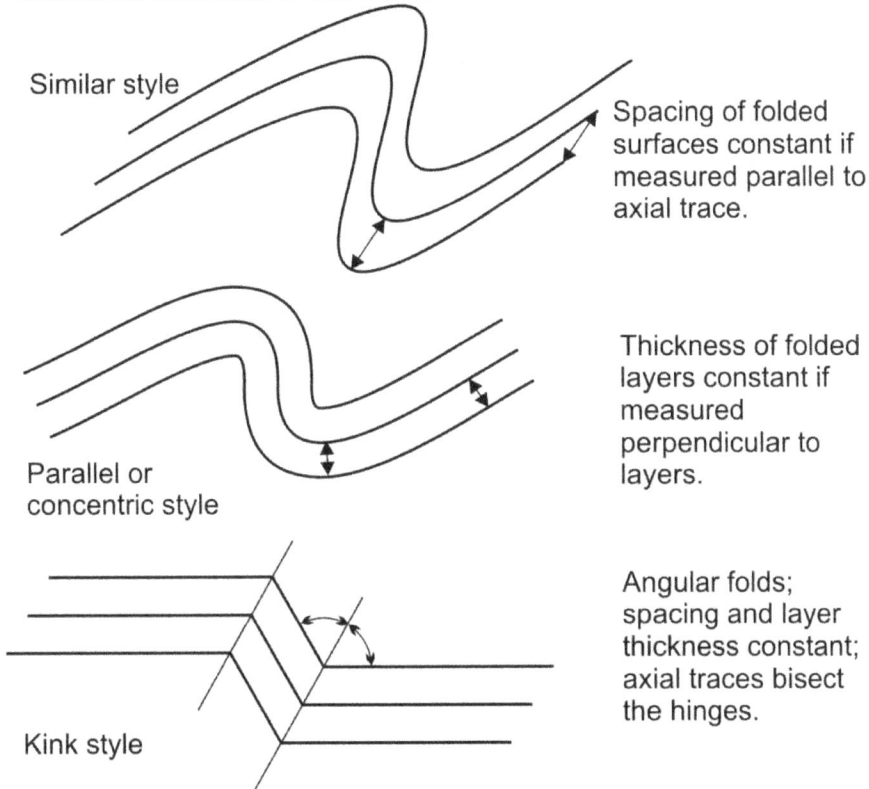

Similar style

Spacing of folded surfaces constant if measured parallel to axial trace.

Parallel or concentric style

Thickness of folded layers constant if measured perpendicular to layers.

Kink style

Angular folds; spacing and layer thickness constant; axial traces bisect the hinges.

Figure 1. Diagrams illustrating idealized characteristics of some common fold styles.

Techniques for folds in profile

In the first part of this lab, you will explore two techniques for constructing cross-sections through subhorizontal folds that are useful in preliminary work on thrust-related folds.

Parallel folds are also sometimes called concentric folds, because the curved arcs are parallel, and are centred on common centres. When a stack of layers exhibits this special geometry, it is possible to make predictions of the continuation of structures at depth that assist with cross-section construction. We will use the **Busk** or **Arc**construction to reconstruct the folds. This construction assumes folds are cylindrical and horizontal folds and the layers have concentric geometry. There is also a helpful tutorial on the worldwide web at the website of Steven Dutch (University of Wisconsin):

⊙ Arc method: https://stevedutch.net/Structge/SL161ArcMethod.htm

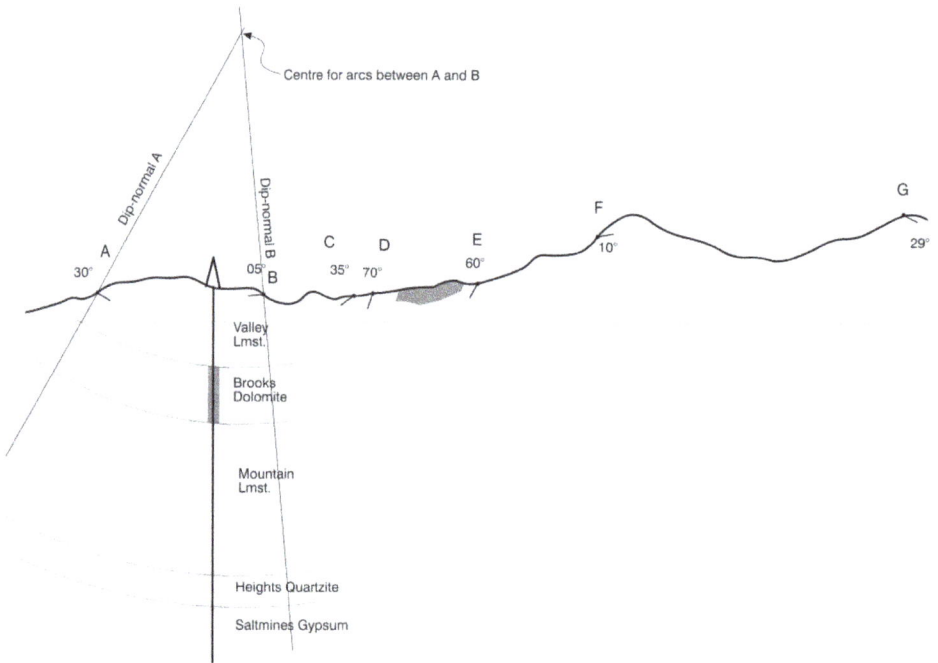

Figure 2. Busk construction, step 1.

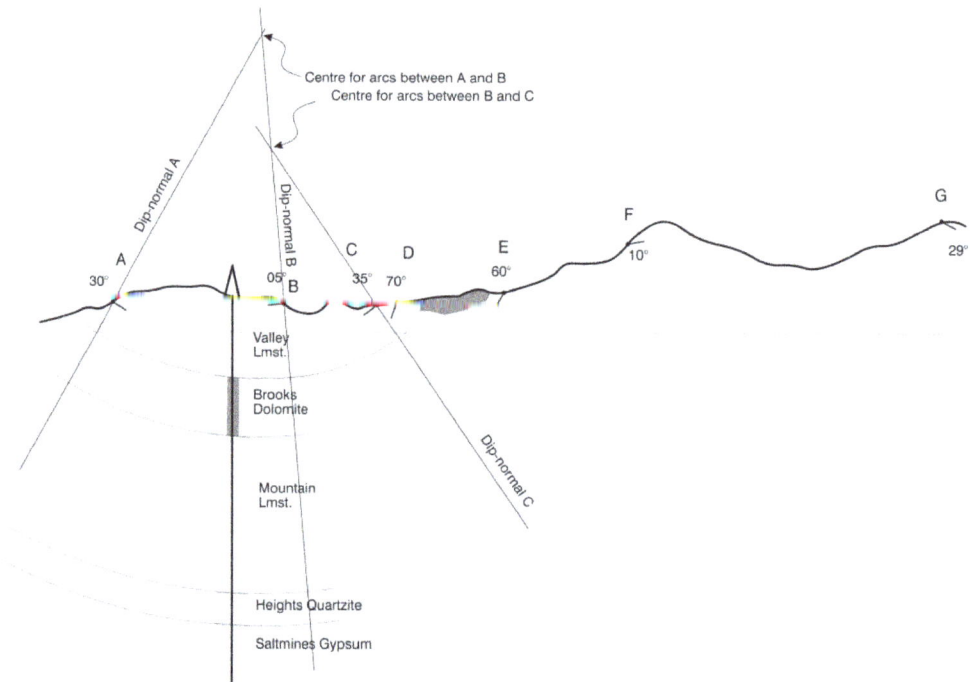

Figure 3. Busk construction, step 2.

In some fold and thrust belts, it has been shown that instead of displaying smooth arcs like those postulated in the Busk construction, folds display straight limbs and angular hinges, like kink folds. This observation leads to the **kink construction**, in which folds are assumed to have straight limbs and angular hinges. Because beds have the same thickness on opposite limbs of each fold, the axial surfaces exactly bisect the inter-limb angles. In the kink construction, rounded fold hinges are approximated by multiple angular folds. Although kink folds depart from parallel geometry at the hinges, these hinge regions are very narrow; as a result, sometimes the kink and Busk constructions give rather similar results. There is also a helpful tutorial on the worldwide web at the website of Steven Dutch (University of Wisconsin). Note, however, that the tutorial makes different assumptions about the locations of fold hinges, compared with the exercise in the assignment, where fold hinges are explicitly located.

- ◉ Kink Method: stevedutch.net/...uctge/SL162KinkMethod.htm

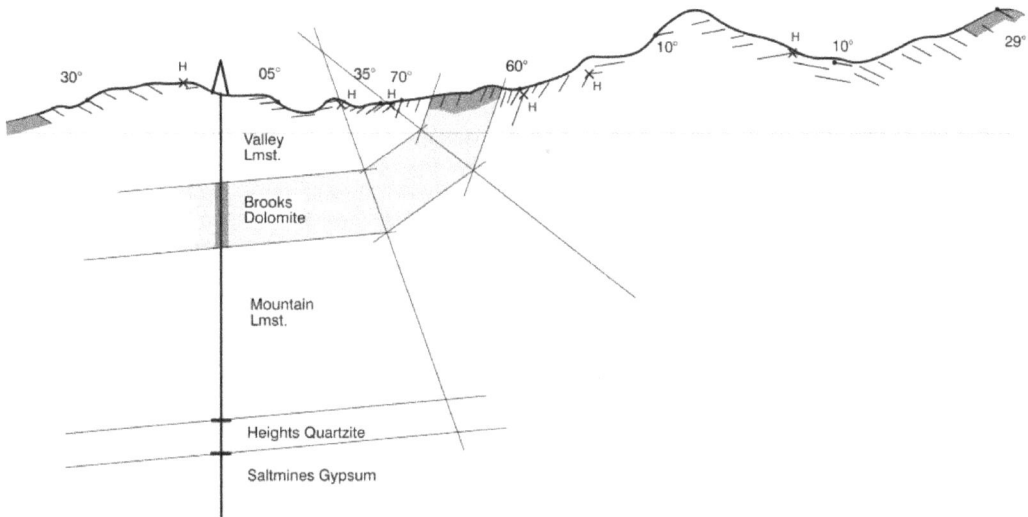

Figure 4. Kink construction, step 2.

Assignment

1. *Busk construction for parallel folds in profile cross-section
- ◉ You are provided with a cross-section through an area of horizontal cylindrical folds, with subsurface information from a single borehole. The scale is 1:10,000, without vertical exaggeration. Dip measurements are shown at intervals at ground level. Make a cross-section using the Busk, or arc construction.
- ◉ To execute the Busk construction, draw a line perpendicular to the dip at each point where the dip was measured. (Use a protractor and the numerical value of dip from the section.) These lines are called dip normals. Give them letters A, B, C, etc.

⊙ Find a place where there is stratigraphic information located between two adjacent dip normals e.g. between A and B. Use a compass with its point placed on the intersection of the bedding normals to extend the stratigraphic boundaries as far as those normals.

⊙ Continue with the next pair of dip normals, using a compass to extend the boundaries, until the section is complete. You may assume that the strata beyond the right-hand end of the section dip at a constant 29°.

⊙ The borehole is extended and encounters the base of the Saltmines Gypsum, which is 450 m thick. Add the base of the Gypsum to the cross section. Notice how the assumption of parallel folding breaks down at depth. (You may find it easiest to start at both ends of the section, measuring the 450 m of gypsum perpendicular to bedding in each case, and working toward the 'difficult' part in the centre.) Why would you expect the gypsum to behave differently?

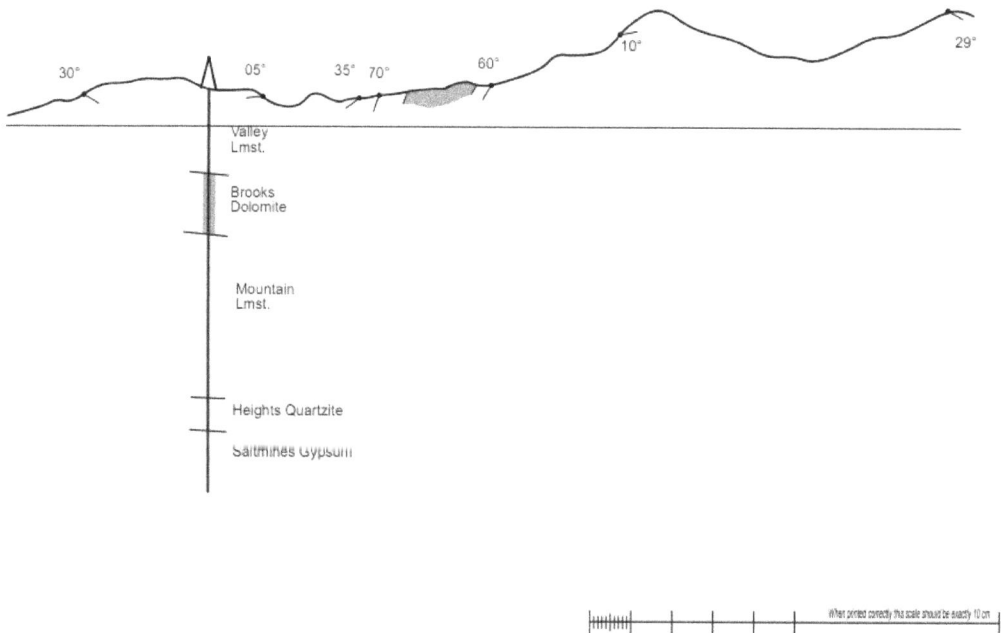

Lab 10. Question 1. Busk Construction [PDF]

2. *Kink construction for angular folds in profile cross-section

You are provided a second cross section that shows some additional information in the area of the previous exercise. A reconnaissance traverse has shown that dip domains are separated by angular folds. The hinges of these folds are marked 'H' with a small x symbol. The new traverse also detected the presence of Brooks Dolomite at both extreme ends of the section.

- At each fold hinge, construct an axial trace that bisects the bedding dips on either side. the easiest way to do this is to add the two dips, and divide by 2; this gives the angle *from vertical* of the bisector. (Be careful if the dips are in opposite directions; you will need to count one as negative and the other as positive.)

- Then draw the stratigraphic boundaries on either side of the axial trace; you should find that the thickness of each unit is unchanged as it crosses the fold. Complete the cross-section as before. Make sure you add the base of the Gypsum as before.

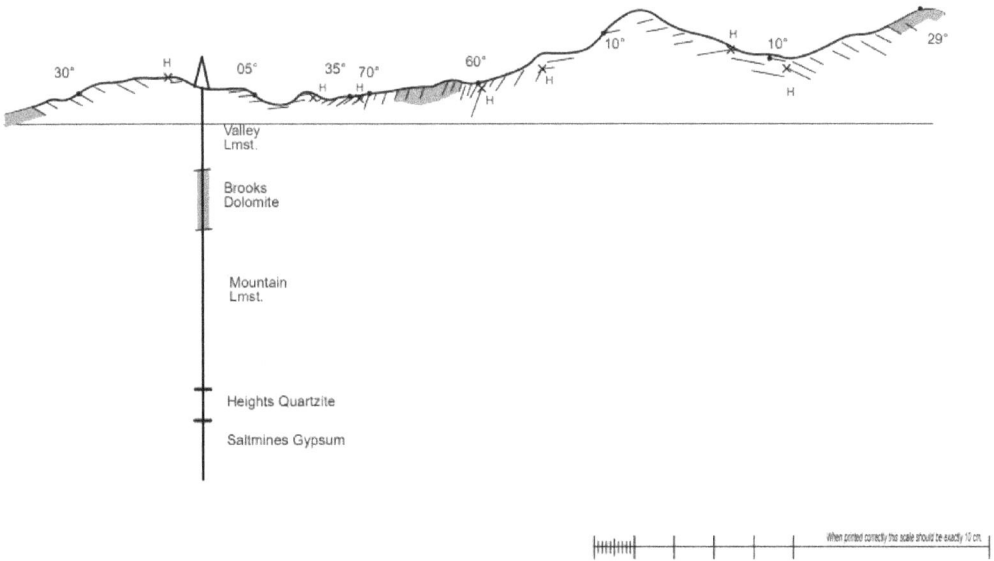

Lab 10. Question 2. Kink Construction [PDF]

3. Thrust belt cross-section

   Map 1 represents a faulted and folded area in a thrust belt. Using the relationships between outcrop shape and topography, structure contours, and fault-bedding intersection lines (cut-off lines), your objective is to determine the structure of the area and construct an accurate cross-section along the marked line.

   a) Preliminary examination of the map

   Look at the outcrop pattern. Both the stratigraphic units and the faults generally have sinuous outcrop patterns that roughly follow contours in parts of the map, suggesting that in these areas both the faults and the stratigraphic units mostly have gentle dips. The repetition of stratigraphic units suggest a dipping fault has caused repetition of stratigraphy. The patterns of faults and outcrops on either side of the main valleys are rough mirror images of each other, suggesting that the valleys have been eroded through structures that are now exposed in both valley sides. Any thrust belt map has a lot of surfaces. Identify the major stratigraphic boundaries and faults with colours as you have done in previous maps. Use red for faults, blues, greens and browns for stratigraphic surfaces.

b) Cut-off lines

A useful first step is to identify cut-off points, where faults intersect stratigraphic units, and connect them, where possible, to draw cut-off lines. Remember that for each surface cut by a fault, there will be a hanging wall cut-off line and a footwall cut-off line. You may notice that corresponding cut-off points have similar elevations on each side of the main valleys. This suggests that the cut-off lines are almost horizontal. Draw possible cut-off lines on the map.

c) Structure contours

Now try to draw structure contours on the dipping surfaces. Note that, because we are in a thrust belt, that surfaces may be repeated, so you may have more than one set of contours on the same surface.

Try to use the minimum number of lines to build the cross-section; however if your map becomes cluttered and difficult to understand, you may wish to use tracing paper for some of the contour sets. Make sure you record the position of the corners of the base map on your tracing paper so that it can easily be located, but *do not attempt to trace all the detail from the map onto your tracing paper*.

d) Cross-section

Project all these lines onto the cross section and complete the most reasonable hypothesis you can for the overall structure on the cross section. Except where you have direct evidence to the contrary, assume a geometry in which layers maintain constant thickness (either a 'Busk' or a 'kink' geometry will do, or a freehand compromise between the two. Your layers should not visibly change thickness across the section.)

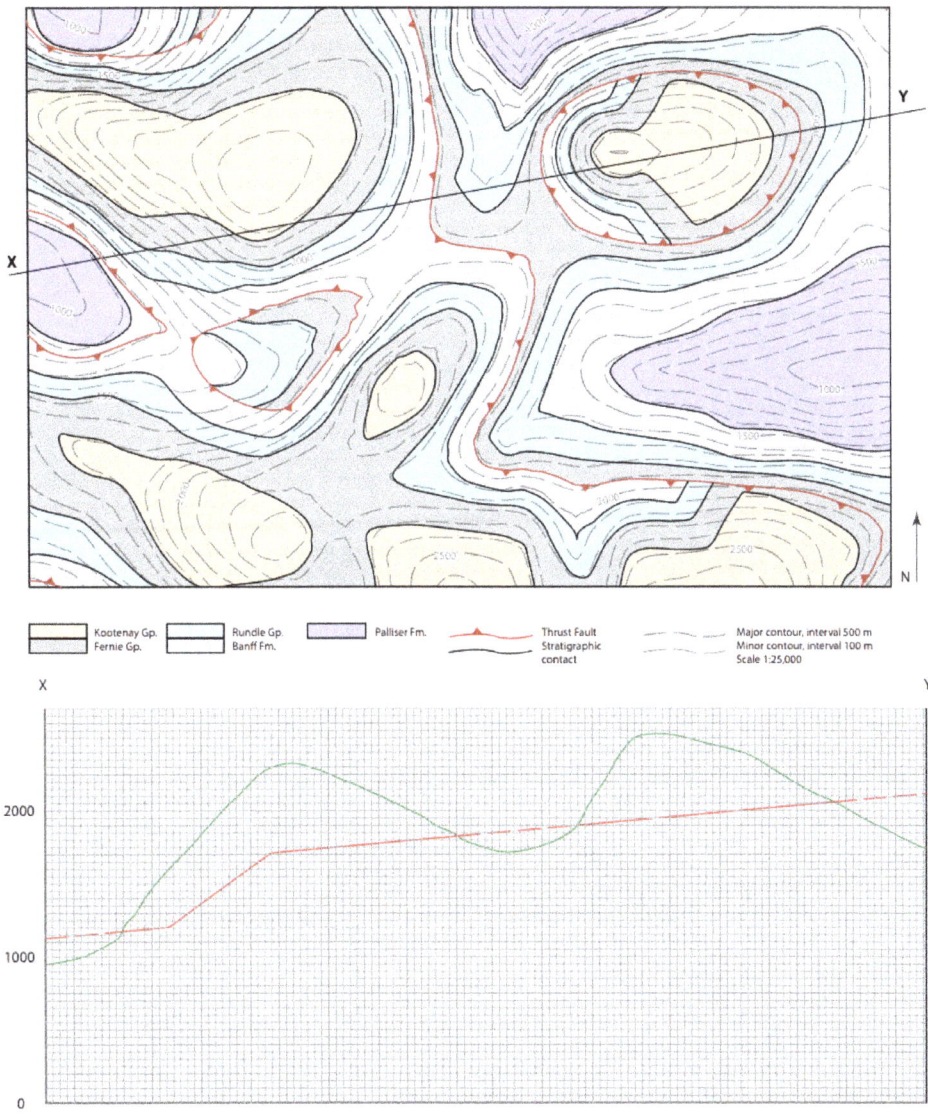

Lab 10. Map 1. Fold and thrust belt (revised version) [PDF]

13 Chapter

SHEAR ZONES

DEFINITION AND GEOMETRY

Shear zones and faults

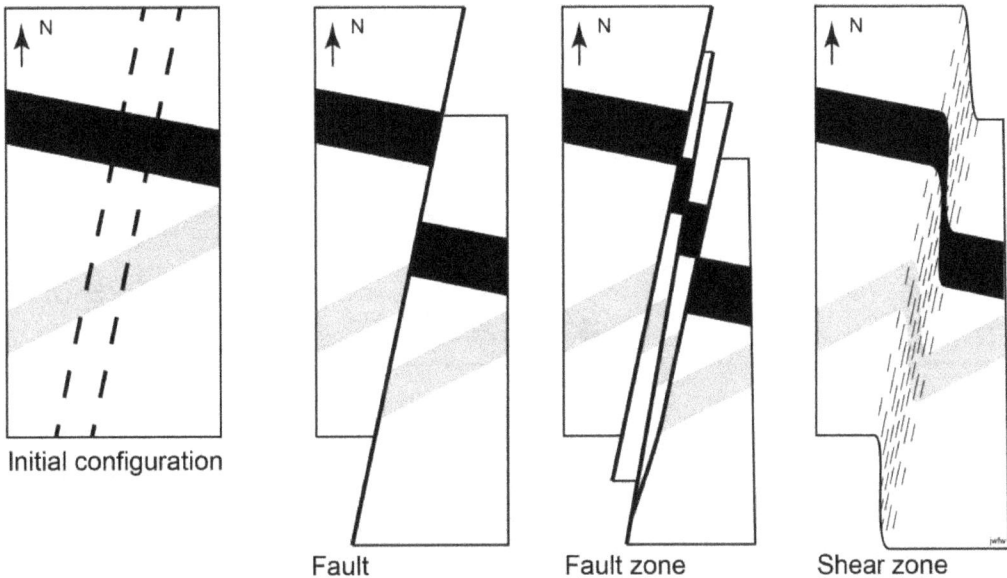

Figure 1. Fault, fault zone, shear zone.

Shear zones are regions of significant ductile deformation that are narrow in comparison to their lateral extent. Shear zones, similar to faults, generally exhibit displacements of older structures; however, they do not possess continuous brittle fractures as faults do. In practice, faults and shear zones are intricately connected. Numerous significant structures that manifest as faults at the Earth's surface likely interconnect with ductile shear zones at depth, and in the transitional region, it is typical to encounter composite zones exhibiting a blend of brittle fracture and ductile movement. At map scale, shear zones can resemble faults and exhibit identical geometric relationships (offset, separation, throw, heave, etc.). This section will not reiterate those definitions. Ductile shear zones are characterized by planar and linear fabrics at both outcrop and microscopic scales. Reduction in grain size is also prevalent. Initially, the metamorphic rock type resulting from this grain-size reduction was believed to be a consequence of brittle grinding (cataclasis), leading to its designation as mylonite, derived from a Greek term for milling flour. The tiny grain size is a consequence of dynamic recrystallization, which involves the disintegration of original mineral grains and the production of new ones due to the accumulation of flaws from severe plastic deformation.

Shear zone kinematics

In instances when the kinematics of shear zones can be ascertained, the majority exhibit a significant degree of rotational deformation, and many are found to have experienced progressive simple shear.Nonetheless, the strain is frequently heterogeneous, with various regions of the shear zone exhibiting distinct aspects of the strain history, a process referred to as strain partitioning..

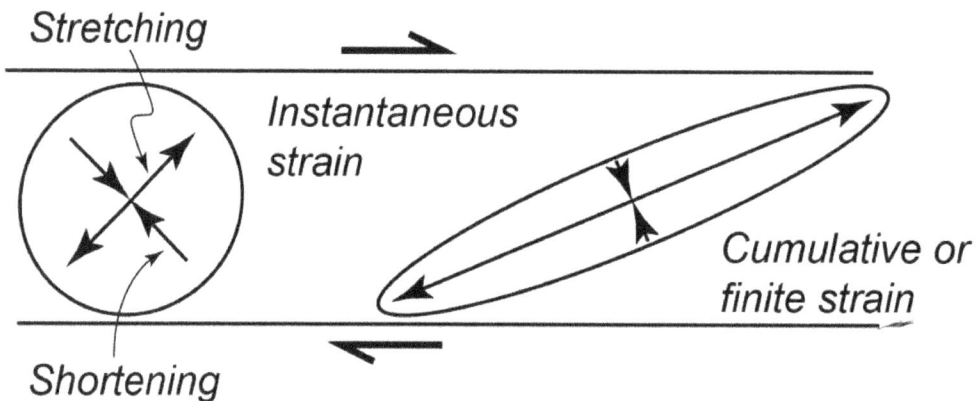

Figure 2. Incremental and cumulative strain in a homogeneous simple shear zone.

The typical geometry of a shear zone is therefore a band of **heterogeneous simple shear**, across which previous structures are offset.

The overall strain inside a zone of ideal simple shear varies. The *incremental strain* typically shows strain axes at about 45° to the shear zone boundary, but the *finite strain* shows the effects of rotation: with increasing deformation the shape of the strain ellipse becomes more and more extreme and the extension direction (X or S_1) rotates closer and closer to the shear zone boundary. Although the extension direction never quite becomes parallel to the shear zone boundary, in cases of extreme deformation, the difference in direction may be almost imperceptible.

Figure 3. Simulation of heterogeneous simple shear applied to a grid of markers. Fabric produced as strain becomes more intense as it curves toward the centre of the shear zone.

FABRICS

Simple sigmoidal foliation patterns

The most basic pattern of foliation in a shear zone is known as **sigmoidal oblique foliation**. This is the fabric that is produced by the flattening of particles or domains (clasts, mineral grains, etc.)in the **protolith** (the original rock), when heterogeneous simple shear takes place.

In the least deformed, marginal parts of the shear zone, the foliation is weak, and oriented at around 45° to the shear zone. Towards the centre of the shear zone the foliation intensifies with increasing strain, and it curves so that it is near-parallel to the overall shear zone. The sense of rotation of the foliation with increasing strain (whether clockwise or counter-clockwise) shows the sense of shear on the zone.

Lineations

Associated with sigmoidal foliation, the stretched particles of the protolith typically are extended and define a **stretching lineation** which gives a streaky appearance to the foliation surface. The orientation of this lineation, when visible, and its rake in the foliation surface, are important quantities to measure in the field description of a shear zone.

C-S fabrics

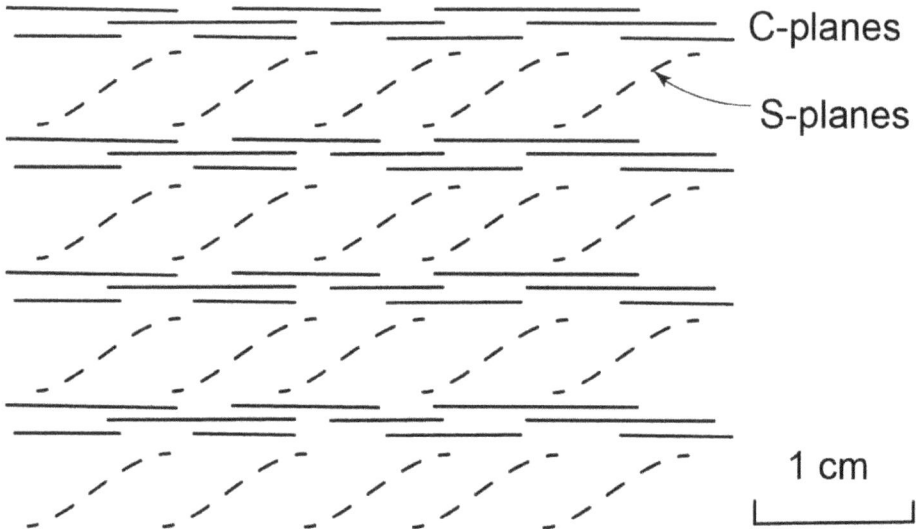

Figure 4. C-S fabric.

In some shear zones, strain is obviously *partitioned* such that zones of intense shear, and grain size reduction, alternate with zones of less intense strain, where the foliation is more oblique and the grain size is coarser. This configuration is called **C-S foliation** (sometimes S-C). The **C-planes** (where 'C' stands for "cisaillement" in French, meaning shearing) are the surfaces that are closer to the shear zone boundary and represent the most deformed bands. The **S-planes** (for "schistosité") represent the less-deformed zones and may be oriented up to 45° from the C-planes and the shear zone boundary. The sense of rotation from S to C shows the sense of shear on the overall shear zone.

Shear bands

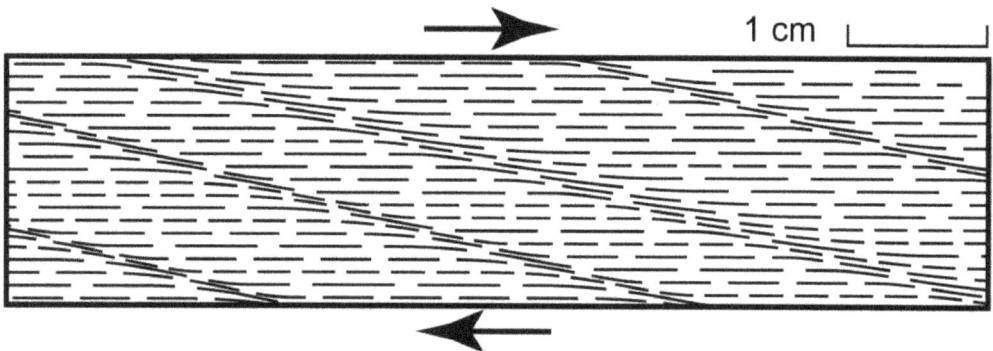

Figure 5. Diagram of shear-band foliation.

Shear band foliations are geometrically very similar to C-S foliations. They form later in the history of a shear zone, typically after a strong foliation has been developed throughout. Shear bands take the form of a new set of smaller shear zones that offset the original foliation. In a dextral shear zone they are clockwise from the original C-planes; in a sinistral shear zone the orientation is counter-clockwise. Shear bands are also known by the names **"extensional crenulation cleavage"** and **"C-prime (C›) foliation"**.

MINERAL STRUCTURE OF SHEAR ZONES

It's possible to learn what goes on in mylonite formation by looking at microscopic thin sections from shear zones. In this type of work it's important to collect oriented rock samples. Typically this is done by marking a strike and dip symbol on a surface of the rock, *before the sample is removed* with a hammer. Any surface can be marked: a foliation surface or an arbitrary weathering surface. It's important to note whether the marked surface is a top surface or an overhanging surface. As an additional check it's useful to flag the right-hand-rule strike direction with an arrow. On a top surface, the dip direction will be clockwise from the RHR strike direction; on an overhanging surface, looking up, the dip direction will appear counter-clockwise from the RHR strike direction.

Undulose (undulatory) extinction

Figure 6. Undulose extinction and mortar structure, sheared quartz vein, Cape Chignecto, Nova Scotia. In this view, polarizing plates are above and below the microscope stage (cross polars) so that the orientation of quartz crystals can be distinguished in different shades of grey. In this example, dynamic recrystallization has started to occur at grain boundaries, but the grain cores show strongly undulose extinction when viewed between crossed polars. Field of view is approximately 1 cm.

On approaching a shear zone, the first thing that is noted is an increase in the intensity of undulose (undulatory) extinction when the thin section is viewed between crossed polars. This is caused by an accumulation of defects (**dislocations**) in the crystal lattice.

Mortar structure

With increasing deformation, new tiny crystals appear along the grain boundaries. These are believed to form because the accumulation of dislocations makes the original crystal unstable, so solid-state recrystallization occurs producing new undeformed crystal lattices. This process is called **dynamic recrystallization.**

Eventually the recrystallized material surrounds the original crystals, producing what is known as **mortar structure** (because it looks like bricks and mortar) or **core and mantle structure.**

Mylonitic rocks

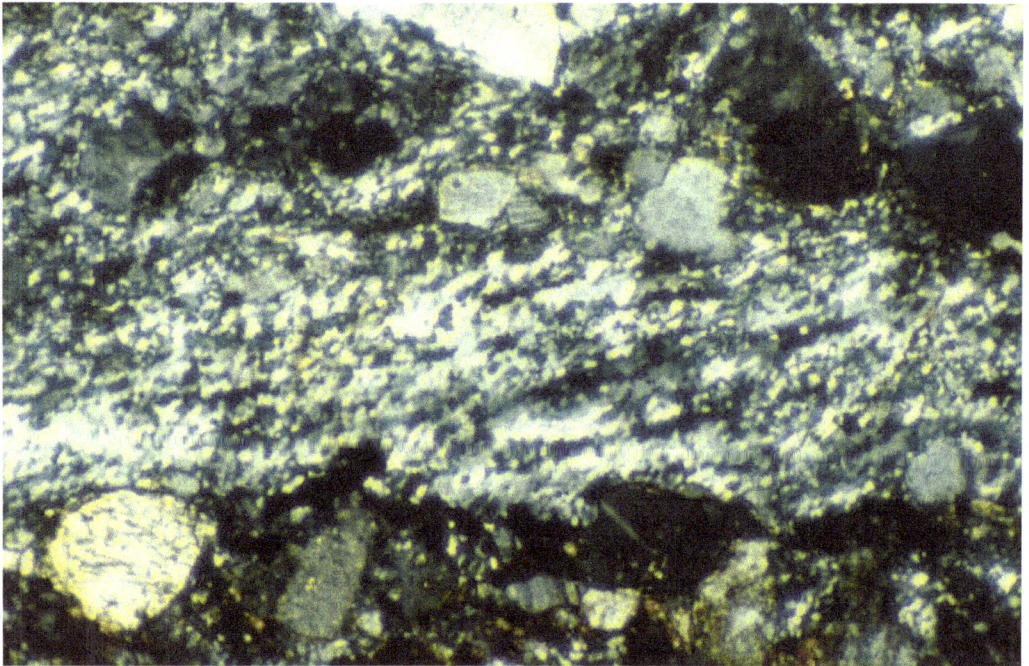

Figure 7. Mylonite, Cape Chignecto, Nova Scotia. In the central part of the slide, the original quartz grains have been reduced to fine-grained mylonite. Field of view is approximately 0.3 cm.

Rocks which contain significant amounts of dynamically crystallized fine-grained material are **mylonitic**. Mylonitic rocks are classified based on the proportion of this fine-grained material relative to remaining fragments of the original rock, known as **porphyroclasts**. (Don't confuse porphyroclasts and porphyroblasts.

Porphyroclasts are large grains in a metamorphic rock that are left over when the rest of the rock has had its grain size reduced by dynamic recrystallization. *Porphyroblasts* are large grains in a metamorphic rock that have grown in a fine-grained matrix, typically as a result of high temperature.)

◉ **Protomylonite**

If the porphyroclasts still make up more than 50% of the rock, then the rock is called a **protomylonite**.

◉ **Mylonite**

Figure 8. Mylonite with porphyroclasts, British Columbia.

True **mylonite** has from 50% to 10% porphyroclasts.

◉ **Ultramylonite**

A rock which is almost entirely dynamically recrystallized, such that there are less than 10% porphyroclasts remaining, is an **ultramylonite**.

Matrix – Porphyroclast Relations

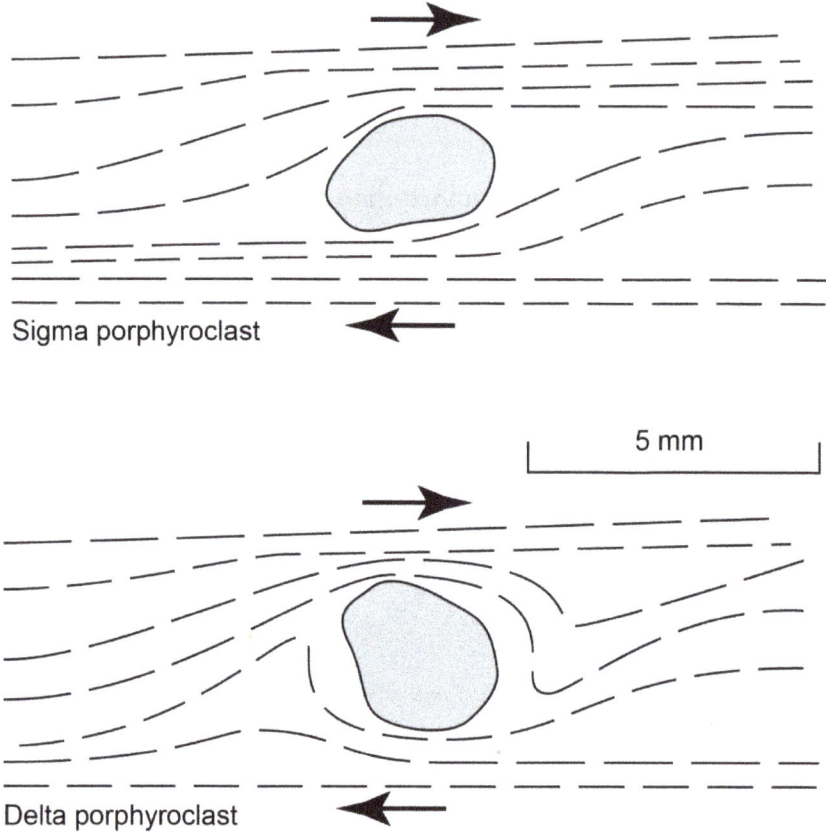

Sigma porphyroclast

5 mm

Delta porphyroclast

Figure 9. Sigma and delta structures.

Because they are less deformed than the surrounding material, porphyroclasts represent sites of strain partitioning, and sometimes give away the sense of shear in a mylonite. There are two distinctive styles.

Sigma structure

The foliation sweeps around a **sigma porphyroclast** in a somewhat rhombic shape, rather like a "small island of S-plane" in a sea of C-plane. Just like C-S fabric it can therefore indicate sense of shear. It is named after the shape of the Greek lowercase letter sigma: σ.

Delta structure

If the porphyroclast is affected more by the rotation component of deformation, it, together with the foliation in the adjoining matrix, may be rolled together into a **delta porphyroclast** named for the Greek letter delta: δ.

Folds in shear zones

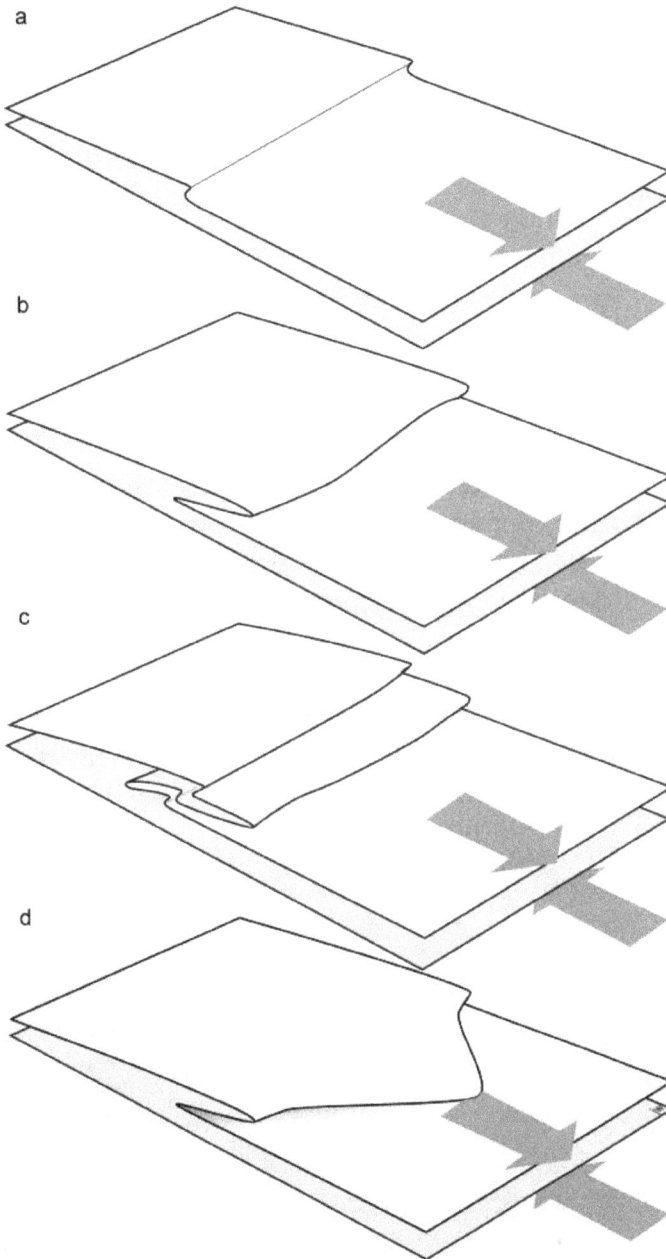

Figure 10. Development of folds in shear zones. In (a), an asymmetric fold is developed, which tightens in (b). In (c), a second asymmetric fold re-folds the first. Alternatively, in (d) the asymmetric fold develops a strongly curved hinge, becoming a sheath fold.

Asymmetric folds

Folds in shear zones typically develop as minor abnormalities in the foliation are intensified by significant strains. Initially, newly formed folds exhibit distinct S or Z asymmetry, contingent upon whether the rotational direction in the shear zone is counter-clockwise or clockwise from the observer's perspective. Nevertheless, the significant stress characteristic of a fully formed shear zone can result in folds becoming isoclinal, with the limbs transposed parallel to the prevailing foliation, making the initial asymmetry challenging to trace.

Refolded folds

Folds in shear zones typically arise when minor abnormalities in the foliation are intensified by significant strains.Initially, newly formed folds exhibit distinct S or Z asymmetry, contingent upon whether the rotational direction in the shear zone is counterclockwise or clockwise from the observer's perspective of the folds. Nevertheless, the significant strain characteristic of a welldeveloped shear zone can result in folds becoming isoclinal, with the limbs transposed parallel to the overall foliation, making the initial asymmetry challenging to trace.

Sheath folds

As folds in shear zones compress, minor irregularities at their hinges are intensified, resulting in fol hinges becoming significantly curled into shapes like to the fingers of a glove.These formations are referred to as sheath folds and are indicative of significantly noncoaxial deformation.

Figure 11. Sheath fold in mylonitic limestone. Old Man's Pond, Newfoundland.

14 Chapter

EXTRATERRESTRIAL IMPACT STRUCTURES

Meteor Crater, Arizona. By Shane.torgerson – Own work, CC BY 3.0, https://commons.wikimedia.org/w/index.php?curid=12188092. This version has been slightly cropped.

OCCURRENCE OF IMPACTS

Extraterrestrial objects intermittently crash the Earth.An inverse relationship exists between the size of an item and its recurrence interval, although the specific figures differ based on whether one accounts for objects entering the atmosphere or solely those that reach the surface.Recurrence intervals for extra-terrestrial impacts of various sizes:

- ⊙ 10 m diameter: approx. 1 yr
- ⊙ 50 m diameter: approx. 1000 yr
- ⊙ 1 km diameter: approx. 500,000 yr
- ⊙ 5 km diameter: approx. every 10 M yr
- ⊙ 10 km diameter: 65 Ma event was probably most recent

The object that exploded in the atmosphere over Russia in February 2013 is estimated to have been ~17 m in diameter.

Objects approaching the Earth may vary in speed between about 11 km s^1 and 72 km s^1, though smaller objects are slowed by drag from the atmosphere.

Meteorite impacts produce some of the highest strain rates on Earth. Typical values of strain rate in the first few microseconds of an impact may be of the order of 10^8 s^1. (Contrast the strain rates during an earthquake, typically 0.1 to 10 s^{-1}. By comparison, time-averaged strain rates in orogens due to plate movement are typically 10^{15} to 10^{12} per second.)

One of the most obvious results of an impact is a roughly circular depression called a **crater**. Major craters have been described from many parts of the Earth. The three largest confirmed examples are:

1. Vredefort Crater in South Africa, diameter ~300 km, formed at 2020 Ma, and now deeply eroded;
2. Sudbury Basin, 250 km diameter, formed at 1849 Ma and subsequently deformed;
3. Chixulub crater, 180 km diameter, 65 Ma.

IMPACT CRATERS

Larger impacts form craters, which have two types of geometry (Fig. 1).

Figure 1. Simple and complex craters.

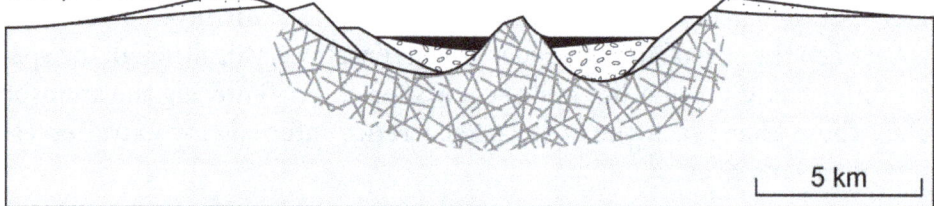

Simple craters

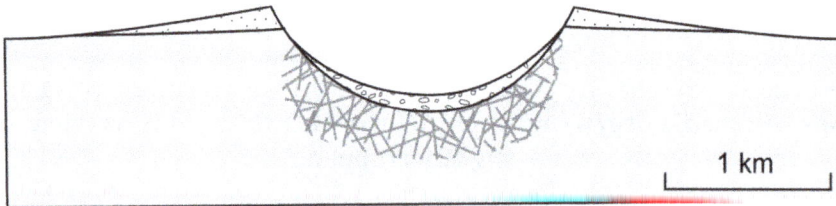

Simple craters are up to about 4.5 km in diameter. They tend to have simple bowl shapes with a raised rim. (On the moon, where gravity is less, simple craters are up to 15 km diameter.)

Complex craters

More extensive craters have intricate geometry.A center elevation is the main distinguishing characteristic of complex craters.The interior of a complex crater's rim generally exhibits a succession of internal terraces, while the exterior may feature concentric ridges and valleys.The floor of a complex crater is frequently quite flat between the central uplift and the terraces.The discrepancies can be elucidated by examining the stages of crater creation, namely the "modification stage." (see below).

IMPACT PROCESSES

Contact and compression

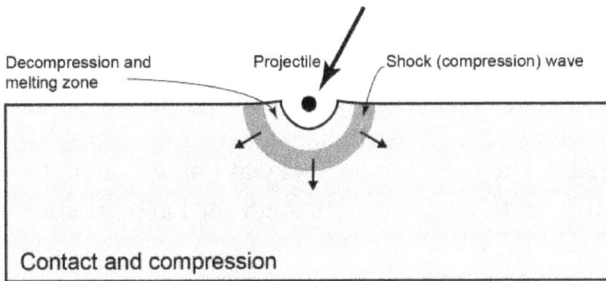

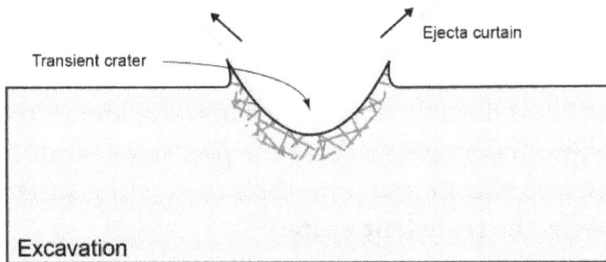

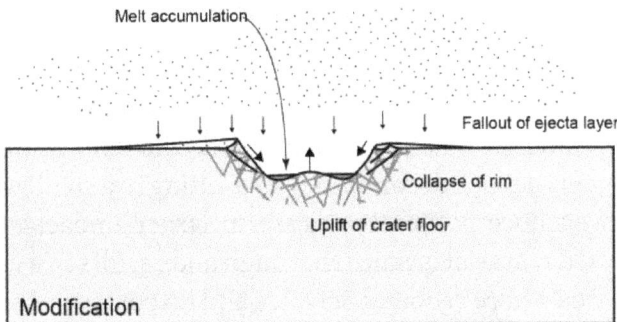

Figure 2. Three stages of crater development.

The first stage in the development of an impact crater is the **contact and compression** stage, beginning when the object first touches the surface. In this stage, the impacting object and the underlying rock are subject to extremely high strain rates of 10^8 per second or higher. These strain rates may produce enormous stresses above 1 TPa (10^{12} Pa), and very high temperatures, sufficient to melt or even vaporize large parts of the impacting body and the rocks in front of it. Rock that remains solid is compressed in a shock wave that initially propagates at a speed comparable to that of the incoming body, typically much faster than normal seismic waves.

As the shock wave passes, rock is highly compressed, which depresses the Earth's surface, starting the process of crater formation. Behind the compressional shock wave, there is a wave of expansion, where the release of elastic energy produces very high temperatures, that melt or vaporize the expanding rock material.

Contact and compression lasts only a few tenths of a second even for a very large impact.

Eventually the speed of the advancing shockwave slows down to seismic velocities (3-7 km s^1) and continues to propagate through the Earth as a seismic wave, during the next stage.

Excavation stage

Initially the compressional wave moves rock downward and outward but as the wave expands material is pushed upwards at the edges of the expanding crater, creating a raised rim. In addition, as the decompression wave passes, material that was formerly compressed downwards expands upward, and its momentum causes it to be thrown up into the atmosphere. These processes result in the excavation of material (ejecta), into the air and onto the surrounding surface. The resulting depression is known as the **transient crater**.

The maximum size of the transient crater marks the end of the excavation phase. The excavation stage lasts several minutes in a large crater.

Modification stage

Crater alteration follows the conclusion of the excavation. The remodeling technique for tiny craters is quite straightforward. The floor of the deep transient crater rebounds when the rock decompresses, resulting in a shallower, bowl-shaped depression. The procedure is more intricate in larger impacts. The rebound of vertically compressed material is more thorough and results in a central elevation. Encircling it is an area where rebound elevates and disperses material, augmenting the rim. Moreover, the rims of these bigger craters are often unstable, and during the modification phase, significant normal faults develop, causing sections of the rim to descend as terraces. Ejecta typically descend back into the crater, and the material may remain in a molten state sufficiently long to accumulate on the crater bottom. Extensive intricate craters typically feature level floors situated between the core uplift and the terraced rim.

Rock products of impact

A variety of impact products can be identified in the ancient record.

Shatter cones

Shatter cones are conical joint surfaces commonly adorned with plumose patterns. The poles to the surfaces of the cones generally correspond to tiny circles when represented on a stereographic projection. The axes of the cones direct towards the place of impact and can assist in identifying this point if their orientations are meticulously measured.

Figure 3. Shatter cones, Charlevoix, Quebec.

Figure 4. Shatter cones, Sudbury, Ontario.

Impact breccia and suevite

Figure 5. Sill of impact breccia, Charlevoix, Quebec.

Impact breccia is generated in regions of increased compression, leading to subsequent expansion. Impact breccia may have voids between the pieces, which could be subsequently filled by vein minerals if hydrothermal fluids permeate, or the voids may be occupied by melt or a matrix of smaller fragments.A finegrained rock exhibiting a breccia texture, composed of a combination of rock fragments, glass, and melt, is referred to as suevite.

Impact melt: pseudotachylite

Figure 6. Impact melt (above) intruding granitoid rock (below), Sudbury, Ontario.

Substantial melting may occur behind the initial shockwave, producing **pseudotachylite**, similar to the pseudotachylites produced by frictional melting along certain faults. However, much larger volumes of pseudotachylite may be found in impact structures than are common along faults.

Pseudotachylites have a much wider range of compositions than those of normal igneous rocks, because they are produced by more or less complete melting of their parent rocks, rather than the progressive and partial melting in most igneous processes.

In the Sudbury structure a thick layer of impact melt underwent fractional crystallization to yield a thick ore body of massive sulphide.

New minerals

The very high pressures encountered in the contact and compression stage may cause new, high pressure minerals to form. The most notable are two high-pressure polymorphs of silica SiO_2. **Coesite** is a monoclinic form of silica in which the silica tetrahedra are linked in a manner similar to feldspars. It requires pressures above 2 GPa to form. **Stishovite** is a very dense phase in which the silicon is in octahedral coordination, surrounded by six oxygen atoms in a tetragonal structure. Stishovite requires pressures above 8 GPa.

Diaplectic glass

Mineral grains subjected to intense and fast compression may undergo structural degradation due to the rupture of atomic bonds. Upon decompression, the atoms may be incapable of diffusing back into a crystal lattice, depending on the temperature, resulting in an isotropic appearance in thin section, akin to volcanic glass. Unlike genuine pseudotachylite, this vitreous substance may retain the shapes and compositions of the original crystals, indicating that it was never completely melted. This substance is referred to as diaplectic glass.

GLOSSARY OF TERMS

Acicular: Needle sh,aped (of mineral grains)

Acive continental margin: A transition or boundary between continental and oceanic lithosphere that coincides with a plate boundary (ridge, trench or transform fault).

Allochthon: A body of rock that has been moved from its original position, usually in the hanging wall of a thrust fault

Allochthonous: Moved from its original position, usually in the hanging wall of a thrust fault

Angular unconformity: An unconformity characterised by an angular discordance (a difference of strike or dip or both) between older strata below and younger strata above.

Anticline: A fold in which the younging direction is away from the inside of the fold.

Antiform: A fold where the limbs dip away from the hinge so that the fold closes upward.

Antitaxial vein: A vein filled with mineral fibres, in which the fibres have grown outwards, toward the walls.

Antithetic Riedel shears (R' shears): Subsidiary fractures that form at about 75° to a fault, with sense of displacement opposite to that of the fault.

Apparent dip: The apparent dip of a surface is its dip measured on a cross-section that is not perpendicular to the strike of the surface. Apparent dip is always less than true dip.

Arc construction: Busk construction

Attitude: Orientation

Autochthon (autochthonous): A body of rock that has not been moved from its original position, usually in the footwall of a thrust fault

Axial planar foliation: Planar fabric that is approximately parallel to the axial surface of a fold.

Axial surface (hinge surface): Imaginary surface that passes through all the hinge lines of a fold

Azimuth: A bearing measured clockwise from north from 0 - 360°.

Bar: A force concentration of one million dynes per square centimetre, or 100,000 Pa

Bathymetric contour: Contours that show the depth of the sea floor below sea level .

Bearing: A direction relative to north.

Beds: Sedimentary layers thicker than 1 cm.

Bioturbation structures (trace fossils): Structures produced by organisms that have disturbed sediment.

Boudin: A structure formed by a layer that has separated into pieces, typically as a result of extension.

Boudinage (pinch-and-swell structure): The process of formation of boudins.

Branch point: A point where one fault meets and joins another.

Breccia: A rock composed of typically angular fragments, typically from a fault. (The term breccia is also sometimes used in sedimentary geology for an angular conglomerate.)

Brittle fracture (brittle failure): Non-recoverable deformation characterized by loss of strength across a surface.

Busk construction: A construction for folds in profile view in which the traces of surfaces are assumed to be concentric arcs of circles. Also known as the arc construction.

C-plane: A surface within a shear zone that is characteristically parallel to the zone boundaries and which shows the most intense deformation; from the French word "cisaillement" meaning shearing.

C-prime (C') foliation: Shear bands that form later in a history of a shear zone after a strong foliation has been developed throughout. They are typically oriented oblique to the shear zone boundaries, with the same sense of rotation (clockwise or counterclockwise) as the overall shear zone.

C-S foliation: Foliation in a shear zone characterized by intensely strained zones roughly parallel to the shear zone (C planes) and less strained zones where the foliation is oblique (S-planes).

Cataclasite: Fragmented material in a fault zone that is sand-sized or smaller.

Chill zone: Typically finer-grained material found near the edge of an igneous intrusion, typically formed by rapid coolling

Chocolate tablet structure: Boudins that are roughly equant in plan view, produced by extension of a layer in all directions

Cleavage refraction: A change in the orientation of cleavage caused by differences in the mechanical properties of layers. Cleavage typically bends so that it is more perpendicular to competent layers, and more parallel to incompetent layers.

Clinometer (inclinometer): A device that measures inclinations.

Coaxial deformation: Deformation in which the strain axes remain fixed to the rock material throughout deformation

Coesite: A monoclinic form of silica in which the silica tetrahedra are linked in a manner similar to feldspars. It requires pressures above 2 GPa to form.

Columnar joints: Joints that separate elongated bodies of rock that are typically polygonal (roughly hexagonal) in cross-section. These typically form perpendicular to the base and top of an igneous flow or sill.

Compaction: Volume loss due to the expulsion of water as rocks are buried.

Compass: A tool which uses the Earth's magnetic field to measure directions parallel to the Earth's surface.

Competence: A general term for resistance to stress.

Complex crater: A crater that show a central uplift and a terraced outer wall.

Compromise boundaries: Approximately planar boundaries produced when mineral grains grow against one another. Compromise boundaries are not controlled by the crystal structure of either mineral grain.

Concentric joints: Joints parallel to the surfaces of an intrusion.

Conjugate shear fractures: Fractures formed when when rocks fail simultaneously along two families of planes approximately 60° apart in orientation.

Contour: A curving line on a map that separates higher values of some quantity from lower values.

Contour interval .: The numerical difference in value between contours on a map

Contour spacing: The horizontal distance between contours recorded on a map.

Convolute lamination: A soft-sediment deformation structure in which laminae are deformed into complex chaotic folds.

Core zone: The central part of a fault, in which brittle deformation has destroyed the continuity of older structures.

Crater: A roughly circular depression formed by a meteorite impact or volcanic activity.

Crenulation cleavage: Planar fabric represented by closely spaced fold axial surfaces. Almost always a second or later generation of fabric.

Crenulation lineation: Linear fabric represented by closely spaced fold hinges.

Crest point: The highest point on the cross-section of an **antiform**.

Crest surface: A surface containing multiple fold crest lines.

Cross bedding: Cross stratification caused by dunes.

Cross lamination: Cross stratification produced by ripples.

Cross stratification: A configuration of laminae produced by migration of bedforms during sedimentation, in which laminae are oriented at an angle to the depositional horizontal.

Culmination point: The highest point on the crest line of a fold.

Cutoff line: The line where a fault intersects an older surface.

Cutoff point: The point on a cross-section where the trace of a fault cuts the trace of an older surface.

Cylindrical fold: A fold with a well defined axis that is everywhere parallel to the folded surface. Parallel cross-sections cut in multiple positions through the fold appear identical to each other.

Damage zone: The outer part of a fault, in which remnants of pre-existing structures can still be traced.

Décollement: A very extensive flat in on a thrust fault, where there has been a large amount of movement.

Deformation: The movements of parts of the Earth's crust relative to each other.

Delta pyroclast: A prophyroclast around which foliation curves in a tightly curved shape, formed as the porphyroclast and adjacent foliated rock are rotated in a shear zone.

Depression point: The lowest point on the trough line of a synform.

Descriptive (geometric) analysis: Analysis comprising records of the positions, orientations, sizes, and shapes of structures that exist in the Earth's crust at the present day.

Detachment fold: A fold produced when wall rocks accommodate variations in slip between different parts of a fault.

Deviatoric stress: The state of stress obtained when the mean stress is subtracted from each of the principal stresses. The deviatoric stress is the part of the stress system that acts to change shape.

Dextral: Right-lateral movement: relative to an observer looking toward a fault or shear zone, the far side appears to have moved right.

Diaplectic glass: Glass generated by the action of a shockwave on common minerals.

Differential stress: The difference between the largest and smallest principal stress.

Differentiated crenulation cleavage: Crenulation cleavage combined with pressure solution cleavage, producing a foliation that is characterized both by folds and by parallel domains of different composition.

Dike: Dyke (U.S. spelling)

Dilation: Change in size (area or volume)

Dip: The value of the inclination of a plane. Represents the inclination of steepest line that can be drawn on the plane.

Dip direction: The azimuth of the steepest inclined line that can be drawn on the plane.

Dip separation: The distance along a fault, measured in the direction of fault dip, between two cutoff lines of the same severed surface.

Dip-slip fault: A fault in which the component of slip parallel to the fault dip exceeds that parallel to the fault strike.

Disconformity: An unconformity between strata of different ages but identical orientation.

Discordance: A difference of orientation between planes: difference of strike or dip or both.

Disharmonic fold: A fold where the hinges and limbs do not match with those in adjacent layers..

Dislocations: A linear defect in the structure of a crystal where the atoms of the crystal structure are out of alignment.

Distortion: Change in shape

Domain: A region within a rock that has a distinctive composition or texture.

Drag fold: A fold adjacent to a fault in which layers are bent in the direction of movement of the opposite wall. Most 'drag' folds probably form during fault propagation and not by 'drag' after the fault has formed.

Ductile: Flow of material in the solid state, typically without fracture.

Duplex: A structure in which imbricate inclined thrust faults merge into flat faults both upward and downward.

Dyke: A minor intrusion where magma has filled a crack that is discordant to layers in the surrounding rocks. (Spelled 'Dike' in the U.S.A.)

Dynamic analysis: Structural methods that involve the determination of force, stress, strength, or energy involved in deformation.

Dynamic recrystallisation: Breakdown of original mineral grains and growth of new mineral grains as a result of extreme ductile deformation.

Earthquake: Rapid relative motion of parts of the solid Earth due to brittle failure on a fault.

Effective stress: Overall stress minus the fluid pressure.

Elastic: A stress-strain relationship in which stress and strain are proportional, and the strain is recoverable.

En echelon: An arrangement of linear or planar structures where each structure offset from adjacent structures in a consistent sense that is oblique to the structures themselves. (Literally: as a squadron of cavalry.)

Endocontact: The region within an intrusion where it is affected by contact with the host rock.

Enveloping surface: A surface tangent to multiple folds in a layer.

Equal angle projection: A spherical projection in which angles are preserved. Also known as a sterographic projection. Great and small circles project as circular arcs. Constructed using a Wulff net.

Equal area projection: A spherical projection in which area is preserved.Great and small circles project as complex, non-circular arcs. Constructed using a Schmidt net.

Exocontact: The region in the host rock (country rock) affected by contact with an intrusion.

Extension: A measure of longitudinal strain equal to the fractional change in length.

Extension fracture: A fracture where rock masses on either side of a fracture have moved apart slightly, without significant movement parallel to the fracture.

Extrapolation: The process of estimating values outside the range of a data set.

Fabric: Any penetrative structure that gives a rock different properties in one direction relative to another.

Fabric element: Features within a rock that are aligned to give the rock a fabric.

Facing direction: Direction of younging of strata in a fold axial surface.

Fault: A fracture showing significant displacement of one wall relative to the other, parallel to the fracture plane.

Fault bend fold: A fold produced by movement of a curved fault, in which movement of the fault caused bending of the hanging wall, footwall, or both.

Fault cutoff: Intersection line between a fault plane and an older planar rock unit.

Fault propagation fold: A fold that forms at the propagating tip of a fault, marking a region where there are rapid changes in both the dip and the slip of the fault.

Fault regime: A tectonic environment in the Earth's upper crust characterized by the orientation of the principal stresses.

Fault scarp: A steep topographic slope marking a fault, where one wall of the fault forms higher ground than the other.

Fault tip: The linear boundary at the edge of a fault surface, separating faulted from unfaulted rock.

Fault zone: A set of sub-parallel faults that may branch and joint along strike or down dip.

Fenster (tectonic window): An area of footwall on a geologic map that is entirely surrounded by hanging wall.

Fibrous vein fill: Mineral grains within a vein that are strongly elongated, typically tracking the direction of movement as the fracture opened.

Finite deformation: The total deformation that a part of the Earth has undergone, between its original configuration and the present-day state.

Finite strain: The total strain that a part of the Earth has undergone, between its original configuration and the present-day state.

Fissility: The property of splitting easily parallel to a plane, usually the result of sedimentary compaction.

Flame structure: A narrow, pointed soft-sediment deformation structure consisting of mud forced up into overlying sand.

Flat: In fault terminology, a region where a fault is parallel to layers in the wall rocks.

Flattening foliation: A foliation defined by tabular domains, produced by shortening.

Floor thrust: The lower flat of a thrust duplex structure.

Fluidise (Fluidize): A process whereby previously stable sediment is behaves as a fluid when upward-flowing water passes through it from below.

Flute: A scoop-like depression in mud that becomes less distinct in a down-current direction.

Flute cast (Flute mold): A sedimentary structure formed when a flute is filled by sand and preserved in relief on the base of a sandstone bed.

Fold axis: The orientation of a line that is everywhere parallel to a folded surface. The fold axis is the direction of the hinge, crest, trough, and inflection lines.

Fold hinge: A line of maximum (tightest) curvature on a folded surface.

Fold interference pattern: A pattern formed by layers that have been folded twice during their deformation history.

Folding line construction: A construction in which differently dipping surfaces are rotated about their line of intersection in order to represent them on a plane sheet of paper.

Foliation: Planar fabric

Footwall: The body of rock directly below a fracture.

Foreland basin: A sedimentary basin that forms adjacent to an orogen, and which deepens toward the orogen, typically formed by bending of the lithosphere in response to the weight of the growing orogen.

Formation: The primary unit of mapping in stratified rock; must be mappable, defined by lithological characteristics, have a type section, and be named for a place or geographical feature.

Fracture: A surface produced by brittle failure at some point in the history of a rock.

Fracture tip: The linear boundary at the edge of a fracture surface, separating fractured from unfractured rock.

Geologic map: A map showing the extent of, and boundaries between, different units of rock.

Geological boundary: A surface in 3D space, or a line on a 2D map, where one type of rock contacts another.

Glacial striae: Lineation on the Earth's surface formed by abrasion during flow of ice, which reveals the direction of ice movement.

Gneissic banding: Foliation characterised by layered domains of different composition, and the parallel preferred orientation of minerals, typicall of high-grade metamorphic rocks.

Gouge: Clay-rich fault material, typically produced by faulting of fine-grained sedimentary rocks.

Graben: A block between two normal faults that dip towards each other. The rocks in a graben are offset downward relative to the rocks on either side.

Graded bed: A bed, usually with a sharp base, in which grain-size becomes finer toward the top.

Gravity regime: A state of stress in which the maximum principal stress is vertical. Normal faults are common.

Great circle: A circular line that divides a sphere into two precisely equal parts.

Groove: A sedimentary structure formed when an object is dragged by a current across a sediment surface.

Groove cast (Groove mold): A sedimentary structure formed when a groove is filled by sand and preserved as a mold on the base a bed.

Group: A lithostratigraphic unit consisting of several formations.

Hackles: Feather-like striae radiating from a central point or line on a fracture. Part of plume structure.

Half-graben: A tilted block bounded by a normal fault on one side, along which the block is offset downward.

Hanging wall: The body of rock immediately above a fracture.

Harmonic folds: Folds in which each layer is folded in step with the adjacent layers, so that hinge points can be matched between layers

Heave: The horizontal component of dip separation.

Heterogeneous strain: Strain that is not the same everywhere within a rock. Straight lines may become bent and parallel lines may becomeg non-parallel.

Hinge: A line or point of maximum (tightest) curvature on a folded surface.

Hinterland: The region adjacent to a thrust belt away from which the thrust sheets appear to have moved.

Homogeneous strain: Strain that is the same everywhere within a rock. Straight lines remains straight, parallel lines remain parallel.

Horst: A block between two normal faults that dip away from each other. The rocks in a graben are offset upward relative to the rocks on either side.

Ideal plastic deformation: A ductile behaviour in which a material shows no deformation a certain stress (yield stress) is reached, and then deforms rapidly so as to prevent the stress from rising futher.

Imbricate fan: A configuration of multiple listric faults that branch upward from a single flat.

Impact breccia: Breccia resulting from a meteorite striking the Earth.

Inclination: An angle of slope measured downward relative to horizontal.

Incremental strain: The strain that occurred during a small interval in the history of deformation.

Infinitesimal strain: The strain that was occurring a single instant during the history of deformation. (The limit of incremental deformation as the time interval tends to zero.)

Inflection line: A line on a folded surface where the curvature is zero, typically marking a change from convex-up to convex-down, or from antiform to synform.

Inflection surface: A surface containing multiple inflection lines.

Inlier: A region on a geologuc map where older strata are surrounded by younger strata.

Interpolation: The process of estimating intermediate values between the points in a sparse set of data.

Intersection lineation: A lineation defined by the intersection of two differently-oriented foliations.

Invariant feature: A feature that is independent of fold orientation.

Inverse grading: Grain-size variation within a bed in which coarser grains are concentrated at the top.

Isobar: A contour line separating higher pressures from lower; a line joining points of equal pressure.

Isopach: A contour line based on the thickness of a stratified rock unit, separating a thicker from a thiner part of a layer; a line joining points of equal thickness.

Joint: A fracture where the movement parallel to the failure surface is minimal.

Joint set: Many joints in approximately the same orientation.

Joint system: A combination of joint sets, cross-cutting each other in a regular way.

Kinematics: The study of how parts of the Earth moved over geologic time. Include changes in position, orientation, size, and shape that occurred between the formation of the rocks and their present day configuration.

Kink construction: A construction for folds in profile view, in which fold limbs are assumed to be perfectly planar, parallel surfaces, and hinges are perfectly angular

Klippe: A region on a geologic map where the hanging wall of a fault is completely surrounded by footwall.

Lamina (laminae): A layer thinner than 1 cm; plural "laminae".

Layers: Domains that are very extensive and parallel-sided.

Lineation: Linear fabric; fabric defined by elements that are parallel to a line in space.

Liquidize, liquidise: A process whereby previously deposited sediment behaves as a liquid, when the sand grains become separated by fluid

Listric: Describes a fault with dip that decreases with depth.

Lithosphere: The rigid outer part of the Earth that is divided into plates. Comprises the crust and the uppermost part of the mantle.

Lithostatic stress: A state of stress in which all principal stresses are equal, due to the weight of overlying rock.

Lithostratigraphic unit: A named rock layer recognized on the base of its lithological characteristics.

Load structure (load cast): A soft-sediment deformation structure caracterized by rounded bulges, typically on the base of a sandstone bed.

LS tectonites: Rocks with a strong tectonic lineation that lies in the plane of a strong tectonic foliation.

Macroscopic (map-scale) structures: Structures that are too big to see in one view. Must be mapped, or imaged from an aircraft or satellite.

Magnetic contour: A contour line based on characteristics of the Earth's magnetic field; a line that separates stronger magnetic fields from weaker.

Magnetic declination: The azimuth of the Earth's magnetic field.

Map-scale (macroscopic) structures: Structures that are too big to see in one view. Must be mapped, or imaged from an aircraft or satellite.

Mapping: The process of recording and interpreting data on a two-dimensional plan view such as a topographic base map.

Mean stress: The average of the three principal stresses.

Member: A smaller mappable unit recognized within a formation.

Mesoscopic (outcrop-scale) structure: Structures visible in one view at the Earth's surface without optical assistance.

Metamorphic aureole: Baked zone around an intrusion that is often recognizable from changes in texture or mineralogy.

Microbreccia: Fine-grained breccia; breccia with particles 2-4 mm in diameter.

Microscopic structure: Structure that requires optical assistance to be visible.

Mineral lineation: Linear fabric defined by parallel alignment of acicular mineral grains; linear equivalent of slaty cleavage and schistosity.

Mortar structure: Microscopic structure characterized by rims of fine-grained recrystallised material surrounding remnants of original crystals.

Mudcracks: Thin fractures on the surface of beds that thin downward, form by shrinkage of mud as it dries.

Mylonite: Fine-grained rock formed through ductile shearing and dynamic recrystallisation in a shear zone.

Mylonitic lineation: Stretching lineation formed during extreme ductile shearing.

Natural scale: The geometry of a cross-section when the vertical and horizontal scales are equal.

Necking: Process whereby layers start to thin at points of weakness during extension.

Negative flower structure: A zone of faults that steeped downwards and merge into a single fault or shear zone at depth, in which central blocks are offset downwards; typical of transtensional and strike-slip faults.

Neotectonics: The study of recent fault movements.

Newtonian (viscous) deformation: Deformation in which the strain rate is proportional to the stress.

Non-coaxial deformation: Deformation in which the strain axes vary in orientation relative to the rock material over time.

Non-cylindrical fold: A fold in which there is no line that lies parallel to all parts of the surface, so there is no fold axis. Cross-sections intersecting the fold at different points show contrasting geometries.

Non-penetrative fabric: A fabric that is not present everywhere in the rock, when observed at a given scale. Individual fabric planes or lines can be counted and the spaces between them can be discerned.

Non-rigid deformation: Dilation and distortion

Nonconformity: An unconformity formed by the contact between younger sedimentary strata deposited upon an eroded surface of older crystalline rock.

Normal stress: The part of the stress (or traction) on a plane that acts perpendicular to the plane.

Oblate: The shape of an ellipsoid with one axis much shorter than the other two. Informally, a "pancake".

Oblique-slip fault: A fault that shows both significant dip slip and significant strike-slip.

Offset: Difference in the cutoff location of a surface on either side of a fault; an informal term for separation.

Onlap: The relationship of successively younger beds that extend farther geographically onto an unconformity surface, generally produced by progressive burial of topography during transgression.

Orientation: The angles between a structure and a frame of reference (typically defined by the north and vertical directions)

Orogenic belt: Regions where the Earth's lithosphere has undergone shortening as a result of plate movements; usually mountainous, or mountainous in the past.

Orthographic projection: A projection method in which structures are projected perpendicularly onto a sheet of paper, often accompanied by folding line constructions.

Outcrop: A region of exposed rock at the present-day erosion surface.

Outcrop map: A map on which observations of exposed rock types and structures are recorded.

Outcrop-scale (Mesoscopic) structure: Structure visible in one view at the Earth's surface without optical assistance.

Outlier: A region on a geologic map where younger strata are completely surrounded by older strata.

Overstep: A relationship involving strata below an unconformity, that describes the way the younger succession rests on a variety of units in the lower succession.

Paleogeologic map:

A map of subcrop units below and unconformity; the geologic map that would be produced if all rocks above an unconformity were stripped away.

Paraconformity: A disconformity where the only evidence for a time-gap is from detailed paleontological investigation.

Parallel fold: A fold in which the thickness of a layer (measured perpendicular to the layer boundaries) is constant.

Pascal: A force concentration of 1 Newton per square metre

Passive continental margin: A transition or boundary between continental and oceanic lithosphere that does not coincide with a plate boundary.

Penetrative fabric: A fabric that is present everywhere in the rock, as far as can be observed at a given scale. Individual fabric planes or lines cannot be counted.

Phyllitic foliation: Penetrative foliation intermediate between slaty cleavage and schistosity. There is no universal agreement on the definition, but typically defined by mineral grains between 0.1 and 1 mm diameter.

Piercing point: The point where an older linear feature intersects a fault.

Pillows: Balloon-like structures formed by rapid chilling of lava erupted under water.

Pitch (rake): Orientation of a line that lies in a plane, measured from the strike direction, within the plane.

Planar surface: A surface with constant orientation (strike and dip); contours are equally spaced, parallel, straight lines.

Plate tectonics: A theory describing the large-scale movements of the the lithosphere using simple mathematical and geometrical methods to describe the movement of rigid plates.

Plume (Plumose) structure: Feather-like and concentric markings on joint surfaces produced during fracture propagation.

Plunge: Value of inclination of a line, measured downward relative to horizontal.

Pole to plane: A line that is perpendicular to a given plane.

Porphyroblast: A large mineral grain in a metamorphic rock that has grown during metamorphism to be larger than the surrounding material.

Porphyroclast: A large mineral grain in a metamorphic rock that is surrounded by finer-grained material produced by grain-size reduction during deformation.

Positive flower structure: An array of downward-steepening faults with similar strike, in which central blocks are offset upwards relative to the surrounding area, typical of transpressional and strike-slip faults.

Pressure: The state of stress in a stationary fluid, or the mean stress.

Pressure solution: A phenomenon whereby minerals in a rock are selectively dissolved in response to stress.

Pressure solution cleavage: A fabric defined by planar domains that originated as a result of pressure solution.

Primary fabric: A fabric that originated as the rock is formed.

Primary structure: A geologic structure formed at the same time as the rock in whichit is found.

Primitive: The outer circle on a stereographic projection. Represents a horizontal plane.

Principal planes of stress: Three mutually perpendicular planes that experience no shear stress.

Principal stresses: Three normal stresses acting along the poles to the principal planes of stress

Principle of Original Horizontality: The principle that most stratified rocks were deposited in layers approximately parallel to the Earth's surface.

Principle of Superposition: The principle that stratified rocks form with the oldest layers at the bottom, and youngest at the top.

Profile plane: The plane perpendicular to a fold axis.

Progressive simple shear: A type of noncoaxial flow in which a single plane (the shear plane) undergoes no rotation, dilation, or distortion; typical of shear zones.

Prolate: The shape of an ellipsoid with one axis much longer than the other two. Informally, a "cigar".

Protolith: The original rock from which a metamorphic rock was formed by the action of heat, pressure, and deformation.

Pseudotachylite: Material melted by heating during fault movement. Is typically dark and very fine grained or glassy.

Pull apart basin: A localised subsiding area formed at a releasing bend on a strike-slip fault.

Pure shear: A special type of coaxial deformation in which there is no dilation and in which the intermediate strain axis stays the same length.

Pure strain: Coaxial deformation: deformation in which the rock material is not rotated relative to the strain axes.

Quadrants: A method of measuring bearings, popular in the United States, where angles are specified clockwise or counterclockwise from N or S, towards either E or W. E.g. S37E

R'-shears: Antithetic Riedel shears.

R-shears: Synthetic Riedel shears.

Radial joint: Joint formed in outside an igneous intrusion. Orientation is approximately perpendicular to the boundary of the intrusion.

Rake (pitch): Orientation of a line that lies in a plane, measured from the strike direction, within the plane.

Ramp; In fault terminology, a region where a fault cuts across layers in the wall rocks.

Refolded folds: Folds that have been subject to a second folding event, producing a fold interference pattern.

Regression: The retreat of the sea from the land surface.

Relay ramp: A region of distortion between the tips of two faults, where slip is transferred from one fault tip to another.

Releasing bend: A bend in a strike-slip fault that produces a component of extension.

Representative fraction: The scale of a map stated as a ratio of the length of a feature on the map to the length of the same feature in the real world.

Restraining bend: A bend in a strike-slip fault that produces a component of shortening.

Reverse drag fold: An obsolete term for a rollover fold or fault bend fold formed in the hanging wall of a listric normal fault.

Ribs: Concentric ridges, approximately perpendicular to hackles on a joint surface. Part of plumose structure.

Riedel fracture (Riedel shear)

Small fractures that develop in response to stresses in fault walls during fault propagation and movement. Rieldel shears are typically oriented at ~15 and ~75° to the main fault.

Right-hand rule: An orientation-measuring convention for planes. When facing the strike direction, the plane dips your right. (Or, dip direction is 90° clockwise from the right-hand-rule strike direction.)

Rods: Linear fabric elements that are elongated, continuous domains.

Rollover anticline: A fault bend fold formed in the hanging wall of a listric normal fault.

Roof thrust: The upper flat of a thrust duplex structure.

Rotation: Change in orientation, typically measured in degrees, about a particular axis of rotation.

Rotational deformation: Deformation during which the strain axes have rotated.

S-plane: Foliation in a shear zone, oriented oblique to the shear zone boundary, representing less deformed regions; from the French word "schistosité".

S-tectonites: A strong secondary planar fabric or foliation. (The S is for 'schistosity'.)

Scalar: A physical quantity that can be represented by a single number.

Schistosity: Penetrative foliation defined by mineral grains coarser than ~1 mm. Coarser-grained version of slaty cleavage.

Secondary structure: A structure formed well after the rock in which it is occurs.

Section balancing: Preparation of two consistent cross-sections showing the geometry before and after deformation. An important part of checking cross-sections through thrust belts.

Separation: The distance between two fault cutoff lines, measured in a specified direction on a fault surface.

Sequence: A package of strata bounded above and below by unconformities.

Shatter cone: A conical joint surface diagnostic of impact structures, often decorated with plumose markings.

Shear fracture: A fracture where the two walls have slid past each other. (More or less synonomous with a fault, but typically used for small-scale features of the Earth and of experimentally deformed rocks.)

Shear stress: The part of the stress (or traction) on a plane that acts parallel to the plane.

Shear zone: The ductile equivalent of a fault zone. A belt of ductile deformation across which movement has caused significant offset between the two sides. Shear zones are typically formed at depths greater than brittle faults.

Sheath fold: A fold having a strongly curved hinge, so that the geometry resembles a finger of a glove. (Sometimes called condom folds.)

Sigma porphyroclast: A structure formed by foliation that sweeps around a porphyroclast forming curved, rhombic shape, indicating the sense of shear.

Sill: A tabular igneous intrusion where magma has intruded parallel to strata in the host rock.

Simple crater

A crater with simple bowl shape and raised rims. Usually less than 5 km in diameter.

Simple shear: A type of non-coaxial deformation in which particles move along lines parallel to a single plane

Sinistral: Left-lateral movement: relative to an observer looking toward a fault or shear zone, the far side appears to have moved left.

Slickenlines: Scratches (striae) or fibres on the fault surface. Indicate direction of net slip.

Slip: The displacement vector of a fault.

Slump structure: A soft sediment structure formed in sediments deposited on a slope that undergo catastrophic slope failure, and move under the influence of gravity. Beds may be tightly folded, boudinaged, or both.

Small circle: lines on a stereonet that reach from east to west, like lines of latitude.

Sole markings: Sedimentary structures, such as flute and groove casts, preserved when coarse sediment is deposited rapidly on a muddy substrate.

Stereogram: A stereographic projection of a geologic structure or structures. A powerful tool for solving geometric problems in structural geology.

Stereonet: A grid of curves, the 3-D equivalent of a protractor. Used to measure angles on a stereographic projection.

Stishovite: A high pressure tetragonal polymorph of quartz, in which silicon is in octahedral coordination, surrounded by six oxygen atoms.

Strain: Change in size and shape (dilation and distortion).

Strain axes: Three mutually perpendicular lines in a strain ellipsoid representing the maximum, minimum, and intermediate stretches; also, the poles to three planes of zero shear strain.

Strain ellipse: The shape of a deformed circle that originally had unit radius.

Strain ellipsoid: The shape of a deformed sphere that originally had unit radius.

Strain partitioning: Deformation in which strain is heterogeneous and different parts of the rock show different parts of the strain history.

Strain rate: Strain per unit time.

Strain ratio: The ratio between the long and short axis of the strain ellipse, a convenient measure of the amount of distortion in two dimensions.

Stratified: Organised in layer (strata) that were originally horizontal.

Stratigraphy: The study of the organization and history of stratified rocks.

Stress: Force concentration or force per unit area; also, the concentration of all the forces acting at a point within the Earth.

Stress axes: The directions of the principal stresses; also, the directions of the poles to the principal planes of stress.

Stress ellipse: The envelope in a plane of all the stresses (tractions) acting at a point. Also, the ellipse having the principal stresses as its axes.

Stress ellipse (2-D) or Stress ellipsoid (3-D): The envelope in 3D of all stresses (tractions) acting at a point. Also, the ellipsoid having the principal stresses as its axes.

Stretch: A measure of longitudinal strain equal to the deformed length divided by the original length.

Stretching lineation: Linear fabric defined by elongated domains, produced by extension.

Strike: Azimuth of a horizontal line that lies in a plane.

Strike separation: The distance along a fault, measured in the direction of fault strike, between two cutoff lines of the same severed surface.

Strike-slip fault: A fault in which the component of slip parallel to the fault strike exceeds that parallel to the fault dip.

Structural geology: The study of structures within the Earth and their origin; in practice, structural geology mainly focusses on secondary structures and the deformation processes that formed them.

Structure contour: A contour based on the elevation of a geological surface, separating higher and lower parts of the surface; a line joining points of equal elevation on a geological surface.

Subcrop: A feature on an unconformity surface formed where the unconformity cuts off (intersects) a older rock unit or surface.

Subcrop limit: A line on an angular unconformity surface making the boundary of an older unit that was partially removed by erosion at the unconformity surface.

Subsurface: The region below the Earth's topographic surface.

Suevite: A fine-grained rock with a breccia texture, consisting of a mixture of rock fragments, glass, and melt, produced during impacts of extra-terrestrial objects.

Surface trace: The line along which a geological surface intersects the topographic surface.

Surface trace (outcrop trace): The intersection of a geological surface with the topographic surface.

Syncline: A fold where the younging direction is towards the centre of the fold.

Synform: A fold where the limbs dip towards the hinge and the fold closes downward.

Syntaxial vein: A vein where repeated cracking has occurred in the centre of the vein and mineral fibres are typically in crystallographic continuity with the grains in the wall rock.

Synthetic Riedel shears (R-shears): Fractures that form at 15° to a fault, with the same sense of displacement as the fault.

Tabular mineral grains: Platy or flake-shaped mineral grains (e.g. mica) that are often aligned to produce a fabric.

Tectonic wedge: A pair of oppositely vergent thrusts that meet in the subsurface.

Tectonic window (Fenster): On a geologic map, an area of footwall that is entirely surrounded by hanging wall.

Tectonics: The mathematical study of structures; commonly applied to large-scale movements of the lithosphere and the structures that these have produced (plate tectonics).

Tensor: A physical quantity that varies in magnitude with orientation, and can be represented by an ellipse or ellipsoid, or by a square matrix of numbers. (Strictly speaking this definition describes a 2nd-order tensor; 1st-order tensors are here referred to as vectors.)

Threading contours: The process of drawing contours separating higher and lower values.

Throw: The vertical component of dip separation.

Thrust regime: A stress regime in which the minimum principal stress is vertical. Reverse faults are common.

Tip line: The linear boundary at the edge of a fracture surface, separating fractured from unfractured rock.

Topographic contours: Contours that show the elevation and shape of the land surface. Contours that separate points of higher and lower elevation.

Topographic surface: The land surface of the Earth.

Trace: The line formed by the intersection of a geologic surface with the topographic surface or a cross-section.

Traction: Force per unit area; also known as stress.

Transform fault: A strike-slip fault that is also a plate boundary.

Transgression: The advance of the sea over the land surface.

Translation: Change in position

Transpression: A combination of strike-slip and shortening.

Transtension: A combination of strike-slip and extension.

Trend: The azimuth of a line, measured in the direction of downward plunge.

Trough point: The lowest point on the trace of a synform.

Trough surface: A surface containing multiple fold trough lines.

Unconformity: An ancient surface of erosion and/or non-deposition that indicates a gap in the stratigraphic record.

Variant feature: A feature in a fold that changes with orientation.

Vector: A physical quantity that has magnitude and direction, and can be represented by an arrow.

Vein: An joint that is filled with minerals, typically deposited from groundwater.

Vergence: The direction in which rocks near the surface have moved relative to rocks deeper down.

Vertical exaggeration: Describes the distortion of a cross-section in which the vertical and horizontal scales are not equal. The vertical exaggeration is the ratio between the length of the representation of a vertical unit line and the representation of a horizontal unit line.

Viscous (Newtonian): A deformation mode in which the strain rate is proportional to the stress.

Wrench regime: A state of stress in which the intermediate principal stress is vertical. Strike-slip faults are common.

Xenolith: A piece of host rock that is surrounded by an intrusion.

Zenith: The highest point on a spherical or curved surface.

INDEX

www.ingramcontent.com/pod-product-compliance
Lightning Source LLC
Chambersburg PA
CBHW061935190326
41458CB00009B/2747

* 9 7 8 1 7 8 7 1 5 3 6 5 3 *